VOLUME FOUR

ADVANCES IN CHEMICAL POLLUTION, ENVIRONMENTAL MANAGEMENT AND PROTECTION

Soil Degradation, Restoration and Management in a Global Change Context

VOLUME FOUR

ADVANCES IN CHEMICAL POLLUTION, ENVIRONMENTAL MANAGEMENT AND PROTECTION

Soil Degradation, Restoration and Management in a Global Change Context

Edited by

PAULO PEREIRA

Environmental Management Laboratory
Mykolas Romeris University, Lithuania

ELSEVIER

ACADEMIC PRESS
An imprint of Elsevier

Academic Press is an imprint of Elsevier
125 London Wall, London EC2Y 5AS, United Kingdom
525 B Street, Suite 1650, San Diego, CA 92101, United States
50 Hampshire Street, 5th Floor, Cambridge, MA 02139, United States
The Boulevard, Langford Lane, Kidlington, Oxford OX5 1GB, United Kingdom

First edition 2019

ISBN: 978-0-12-816415-0
ISSN: 2468-9289

For information on all Academic Press publications visit our website at https://www.elsevier.com/books-and-journals

Publisher: Zoe Kruze
Acquisition Editor: Jason Mitchell
Editorial Project Manager: Shellie Bryant
Production Project Manager: Denny Mansingh
Cover Designer: Miles Hitchen

Typeset by TNQ Technologies

CONTENTS

CONTRIBUTORS

Jose A. Acosta
Sustainable Use, Management and Reclamation of Soil and Water Research Group, Department of Agricultural Engineering, Universidad Politécnica de Cartagena, Cartagena, Colombia

José Arnáez
Área de Geografía Física, Departamento de Ciencias Humanas, Universidad de La Rioja, Edificio Luis Vives Logroño, Spain

Amber M. Bateman
The University of Western Australia, School of Biological Sciences, Crawley, WA, Australia; Kings Park Science, Department of Biodiversity, Conservation and Attractions, Kings Park, WA, Australia

Igor Bogunovic
University of Zagreb Faculty of Agriculture, Department of General Agronomy, Zagreb, Croatia

Loredana Canfora
Consiglio per la Ricerca in Agricoltura e l'Analisi Dell'Economia Agraria, Centro di ricerca Agricoltura e Ambiente, Roma, Italy

Ángel Faz
Sustainable Use, Management and Reclamation of Soil and Water Research Group, Department of Agricultural Engineering, Universidad Politécnica de Cartagena, Cartagena, Colombia

Manuel Pulido Fernández
GeoEnvironmental Research Group, Department of Arts and Territory Sciences, University of Extremadura, Cáceres, Spain

Carla S.S. Ferreira
Research Centre for Natural Resources, Environment and Society (CERNAS), Polytechnic Institute of Coimbra, Agrarian School of Coimbra, Coimbra, Portugal

Marcos Francos
Departamento de Ciencias Históricas y Geográficas, Universidad de Tarapacá, Arica, Chile

María Gabarrón
Sustainable Use, Management and Reclamation of Soil and Water Research Group, Department of Agricultural Engineering, Universidad Politécnica de Cartagena, Cartagena, Colombia

Zahra Kalantari
Stockholm University, Department of Physical Geography and Bolin Centre for Climate Research Stockholm, Sweden

Ivica Kisic
University of Zagreb Faculty of Agriculture, Department of General Agronomy, Zagreb, Croatia

Teodoro Lasanta
Instituto Pirenaico de Ecología, Procesos Geoambientales y Cambio Global, IPE-CSIC, Zaragoza, Spain

Maria Burguet Marimón
Department of Geography, University of Valencia, Valencia, Spain

Silvia Martínez-Martínez
Sustainable Use, Management and Reclamation of Soil and Water Research Group, Department of Agricultural Engineering, Universidad Politécnica de Cartagena, Cartagena, Colombia

Juan F. Martínez-Murillo
Departamento de Geografía, Universidad de Malaga, Malaga, Spain; Instituto de Geomorfología y Suelos (IGSUMA), Universidad de Málaga, Malaga, Spain

Miriam Muñoz-Rojas
The University of Western Australia, School of Biological Sciences, Crawley, WA, Australia; Kings Park Science, Department of Biodiversity, Conservation and Attractions, Kings Park, WA, Australia; UNSW Sydney, Center for Ecosystem Sciences, School of Biological, Earth and Environmental Sciences, Sydney, NSW, Australia

Estela Nadal-Romero
Instituto Pirenaico de Ecología, Procesos Geoambientales y Cambio Global, IPE-CSIC, Zaragoza, Spain

Paulo Pereira
Environmental Management Center, Mykolas Romeris Univeristy, Vilnius, Lithuania

Luca Salvati
Italian Council for Agricultural Research and Economics Arezzo, Italy

Rory P.D. Walsh
Department of Geography, College of Science, Swansea University, Swansea, United Kingdom

Ilaria Zambon
Tuscia University, Department of Agricultural and Forestry Sciences Viterbo, Italy

Raúl Zornoza
Sustainable Use, Management and Reclamation of Soil and Water Research Group, Department of Agricultural Engineering, Universidad Politécnica de Cartagena, Cartagena, Colombia

SERIES EDITOR'S PREFACE

Paulo Pereira is the editor of this volume 4 of *Advances in Chemical Pollution, Environmental Management and Protection* series on "*Soil degradation, restoration and management in a global change context.*" As pointed out by Paulo in his preface he claims that there is no more time to wait to reduce deforestation, soil degradation and adopt as well sustainable agricultural practices. With one-fifth of the world's soils currently degraded, and continuing to degrade at a rate of 5–10 billion hectares annually, the burden and responsibility of addressing soil conservation and managment falls at all levels of our society regardless of class, or geographical origin. In addition to soil degradation we need to face soil erosion and salinization, intensive agricultural practrices, food security and freshwater scarcity, a change of the climate and socioeconomic impacts. Soil contamination by agricultural practices, not only by pesticdes and fertilizers but also also by microplastics is being addressed in the last few years. Residues of microplastics in soil originate from plastic mulching, plastic wastes, soil amendments and irrigation practices. Such type of pollution is considered to be higher than in oceans.

That being said, the United Nations targets by 2030, are directed towards combat desertification, restore degraded land and soil, including land affected by desertification, drought and floods, and strive to achieve a land degradation-neutral world. This is a timely book that clearly describes soil research and management issues by well-known scientists. The book contains six chapters written by experts in this field with a comprehensive list of topics: Soil degradation, processes, future treats and possible solutions, agriculture and grazing environments, abandoned and afforested land, and environments affected by fire.

The target audience of this book is not only soil scientists, but much broader covering undergraduates, graduates, educators, environmental scientists as well as landholders and governments officers involved in agroecosystems, land management and urbanization in general that want to learn more about soil protection.

Finally I am extremely grateful to everyone who contributed to this book. In this way I would especially like to thank Paulo for his great job. I am sure that this book will be of great help to better understand and preserve soil at all levels.

D. Barceló
IDAEA-CSIC and ICRA, Barcelona and Girona, September 27, 2019

PREFACE

Soil is the base of life and human activities. There are many direct and indirect ecosystems services that we receive from soils such as food, carbon storage, water storage and pollutants filtration, flood regulation, clean air and shelter. They are key to climate change mitigation and poverty alleviation.[1] Soil functions and services are crucial for the most important challenge that we have ahead. Meet by 2030 the United Nations Sustainable Development goals[2] and reduce our climate and ecological footprint. There are in numerous publications that highlight the importance of soils to achieve a global sustainability, and this book aims also to be a contribution to be this be achieved. Despite the importance of this medium term goal, measures are need to reduce soil degradation. The coupled impact of land-use and climate change are threatening the ecosystems and soils health.

There is no more time to wait, humanity needs to reduce deforestation, soil degradation and adopt sustainable agricultural practices. Per minute approximately 23 ha of agricultural land is lost as consequence of desertification or drought. 75% of land is degraded and have impacts on the well-being of approximately 3.2 billion persons and contributing importantly to the sixth mass extinction and costs more than 10% of annual gross product in loss of ecosystem services and biodiversity. Per year, we lose 13 million of hectares of forest that is the habitat for 80% of terrestrial species and supports 1.6 billion people. At the current trend by 2050, 95% of the land surface is affected by degradation. This will affect food security and obligate millions to migrate.[3,4] The economic impacts of degradation are unbearable. Land use changes is imposing an annual loss of US$231 billion per year,[5] especially in subsaharian Africa, where the vulnerability to famine and social instability is the highest.

United Nations sustainable goal 15 "Life on Land" aims to "Protect, restore and promote sustainable use of terrestrial ecosystems, sustainability manage forests, combat desertification and halt and reverse land degradation and halt biodiversity loss." Under the target 15.3 this goal aims to ***"By 2030, combat desertification, restore degraded land and soil, including land affected by desertification, drought and floods, and strive to achieve a land degradation-neutral world"***. At the current pace it is estimated that by 2020 is very likely that this target will not be achieved since among other aspects, land degradation and biodiversity is continuing to increase.[6] Our lack of capacity to meet this goal at short term is attributed to several aspects that will be debated and discussed in this book. Population and food demand is increasing at an unprecedented rate. In 2019, the population was 7.7 billion persons and it is estimated that by 2050 it will be 9.7 billion.[7] This rapid increase will have impacts on the expansion of croplands and agriculture intensification. The demand for land and the large exploitation of soil resources are one of the most important causes of soil degradation. Our demand for palm oil and meat is increasing deforestation at an alarming rate. We are becoming an urban society, and in the last 100-50 years, agricultural areas started to be abandoned. This rural exodus affected importantly ecosystems that were extensively managed and were part of a cultural landscape. This abandonment, lead to a high accumulation of biomass, which is one of the causes of the high frequency of high severity fires. The demand mineral resources and the consequent abandonment of these activities have also important implications on soil resources, since mine activities have an important impact on the soil degradation, sometimes with a high cost of restoration.

The aim of this book is to review important aspects soil degradation and restoration practices in a global change context. In the first chapter we have a global overview about soil degradation at international level and in the second the impacts of agriculture and grazing and the current restoration measures will be discussed. The third chapter is focused on the impacts of land abandonment on soil degradation, while the fourth is devoted to understand the impacts of fire on soil properties and degradation. Finally the fifth and the sixth chapters are dedicated to areas where humans have a strong impact on soil degradation, namely mining and urban areas. I strongly believe that this book will be extremely important for the understanding of the human activities impact on soil degradation. The authors of the chapters are well recognized scientists from all around the world (Europe, America and Australia) in the field and share in this compendium many years of research. Last but not the least I would like to thank Prof. Damia Barceló for the opportunity to edit this book. Your guidance and experience are extremely appreciated in my research career. I would like to acknowledge the hard work and patience of Shellie Bryant and Jason Mitchell for the professionalism in handling the submission and support me with many technical questions.

Paulo Pereira
Environmental Management Laboratory
Mykolas Romeris University, Lithuania

Acknowledgment

This edition of this book was supported by the Croatian Science Foundation through the project "Soil erosion and degradation in Croatia" (UIP-2017-05-7834) (SEDCRO)

References

1. Pereira P, Bogunovic I, Munoz-Rojas M, Brevik EC. Soil ecosystem services, sustainability, valuation and management. Curr Opin in Environ Sci Health 2018;5:7–13.
2. Keesstra SD, Bouma J, Wallinga J, Tittonell P, Smith P, Cerda A, Montanarella L, Quinton J, Pachepsky Y, van der Putten W, Bardgett RD, Moolenaar S, Mol G, Jansen B, Fresco LO. The significance of soils and soil science towards the realization of the United Nations Sustainable Development Goals. Soil 2016;2:111–28.
3. United Nations. Promoting the sustainability of terrestrial ecosystems and halting desertification, land degradation and biodiversity loss. Available in: https://wedocs.unep.org/bitstream/handle/20.500.11822/25765/SDG15_Brief.pdf?sequence=1&isAllowed=y.
4. IPBES. The IPBES assessment report on land degradation and restoration. In: Montanarella L, Scholes R, Brainich A, editors. Secretariat of the Intergovernmental Science-Policy Platform on Biodiversity and Ecosystem Services, Bonn, Germany; 2018. p. 744.
5. Nkonya E, Mirzabaev A, von Braun J. Economics of Land Degradation and Improvement – A Global Assessment for Sustainable Development. Springer; 2016686.
6. Sutainable development goal 15. Available in: https://sustainabledevelopment.un.org/sdg15.
7. United Nations. World Population prospects 2019. Highlights. Available in: https://population.un.org/wpp/Publications/Files/WPP2019_Highlights.pdf; 2019.

CHAPTER ONE

To whom the burden of soil degradation and management concerns

Amber M. Bateman[a,b,*] and Miriam Muñoz-Rojas[a,b,c]

[a]The University of Western Australia, School of Biological Sciences, Crawley, WA, Australia
[b]Kings Park Science, Department of Biodiversity, Conservation and Attractions, Kings Park, WA, Australia
[c]UNSW Sydney, Center for Ecosystem Sciences, School of Biological, Earth and Environmental Sciences, Sydney, NSW, Australia
*Corresponding author: E-mail: amber.bateman@research.uwa.edu.au

Contents

Abstract

Soil degradation and conservation is an environmental, social, economical and cultural issue concerning the wellbeing of all humankind. With one-fifth of the world's soils currently degraded, and continuing to degrade at a rate of 5–10 billion hectares annually, the burden and responsibility of addressing soil conservation and managment falls on every individual; farmer, land owner, indigenous community, politician, scientist, worldleader and the general public. The causes of soil degradation are multifaceted

Advances in Chemical Pollution, Environmental Management and Protection, Volume 4
ISSN 2468-9289
https://doi.org/10.1016/bs.apmp.2019.07.001

with natural and anthropogenic drivers of global change creating a reduction in soil function, ecosystem services and land productivity, resulting in the deterioration of the social and economic framework of society. To combat land degradation among other issues, the United Nations General Assembly created 17 Sustainable Development Goals (SDGs) as part of the "2030 Agenda for Sustainable Development". These goals highlight the importance of soil conservation by uniting landholders globally to achieve Zero Net Land Degradation. However, these SDGs are only the first step in identifying the issues associated with land degradation and future targets. Further consideration of soil indicators along with landholder engagement is required to evaluate the progress of sustainable land. Governments and scientists need to nurture collaboration with farmers, create policy that supports and strengthens sustainable land management and increase knowledge sharing at all levels of governance and in the education system. Through coordinated, collaborated and sustainable management of our soils we can create resilient and productive landscapes capable of supporting a growing human population in the context of global change.

Keywords: Global change; Land degradation; Land use and management; Soil function and recovery; Land degradation neutrality; Sustainable development goals; Soil ecosystem services; Stakeholder engagement and education; Farmer perceptions; Government science policy.

1. Introduction

The Earth has undergone and is continuing to experience extreme global change, and with it, humans must change also change. The human population is expected to increase by 37% to reach 9.7 billion by the year 2050.[1] As the human population increases, so will the anthropogenic pressures on soils to provide goods and services such as food and fiber to support society.[2] Thus, the effective understanding and management of this finite resource is crucial if we are to meet global food production demands with an ever-growing population. Although soil science has brought this issue to the forefront globally through the establishment of international soil societies and the integration of soil science into the education system, the continued degradation and erosion of our soils is happening at a concerning rate of 5−10 billion hectares annually.[3−5] The regeneration and recovery of soils, especially those under intensive use through agriculture or extractive operations such as mining, is of the highest priority for the conservation of soils.[6]

At the United Nations Assembly in 2015, a set of global standards were designed to address issues that drastically affected the well-being and livelihoods of humankind, such as large-scale land and soil degradation. These 17 Sustainable Development Goals (SDGs) created as part of the "2030 Agenda for Sustainable Development" strive for a more sustainable future with one aspect of these goals, as highlighted in goal 15.3, seeking to build a land degradation neutral world where food production is intensified on existing cropping and pasture lands under sustainable land management practices.[3,5,7] In addition, these goals aim to restore lands affected by degradation processes such as erosion, contamination or fertility loss, and return them to productive lands that provide ecosystem services. Although the SDGs seek to unite the globe under one mission, they are only designed for world leaders and scientists and not accessible, nor beneficial, to landholders and farmers.[7–9] Appropriate stakeholder engagement is required if we are to achieve goals for the restoration of degraded soils to recover productive lands. The SDGs need to be accessible to landholders, indigenous communities, farmers and miners to achieve localized, on ground development of sustainable land management. Currently, farmers worldwide recognize the impacts associated with soil degradation, such as soil erosion, salinization, pollution, decline in productivity and climate change resilience, but do not have access to technology and the necessary capital required to support the introduction of sustainable land management and regenerative farming practices.[10,11] With increased engagement and collaboration between world leaders, scientists and landholders worldwide, the SDGs could transform from ambitious "pie in the sky" ideals and be broken down into localized targets.[7] Ultimately, soil degradation is a global issue affecting the livelihood of billions of people (directly or indirectly) and the burden of soil degradation and sustainable land management falls on all of humankind regardless of class, wealth or geographical origin.[12]

In this review, we seek to (i) evaluate the current state of soil and land degradation globally, (ii) determine the leading causes for soil degradation, (iii) evaluate the impacts of soil degradation on the social, economic, environmental and political well-being of global society, (iv) establish the role of the people as the true and ultimate stakeholder with regards to the burden of soil degradation and management and, (v) open up discussion surrounding current knowledge and levels of community awareness of, and involvement in, issues of soil degradation.

2. Global soil degradation

The Earth's soils are crucial for the survival of mankind and life on the planet as they are responsible for ecosystem services that sustain society. Soils filter and purify water and air, provide fundamental nutrients and minerals, and support plant life, which is a source of food.[1,13–16] Soil and landscape degradation, in the form of loss or deterioration of soil quality, is one of the most critical challenges facing humankind.[12] An estimated 23% of the Earth's soils are degraded and are continuing to degrade at a rate of 24 billion tonnes annually, costing the global economy an excess of US $42 billion per annum.[3,6,17–19]

The level and extent of global soil degradation varies across continents depending on the anthropogenic pressures placed on soil to deliver services. Hotspots for land degradation include the sub-Saharan Africa, which makes up 13% of the global total area of degraded land, followed by the Indo-China region including Myanmar, Malaysia and Indonesia which make up 6% of the global land degradation area.[20] In India, a large exporter of agricultural products, an estimated 147 million ha of soil are severely degraded as a result of both natural (droughts, landslides and floods etc.) and anthropogenic events (land clearing and poor management).[21] Across Europe, various degradation assessments estimate land degradation of the region to range from 60 to 403 million ha.[22] In Mexico, 4,200 ha of land were abandoned due to poor soil quality between the years 2005–10.[23] In Australia, there are over 50, 000 abandoned mine sites with no future plans for restoration.[24,25] Estimates of abandoned mines across the globe exceed one million exacerbating the contribution of the mining sector on soil degradation.[24] Arid lands are experiencing extensive levels of soil degradation with up to one-fifth of drylands currently degraded.[26,27] These dryland systems make up to 40% of the Earths land surface and are responsible for a large portion of the food that feeds the globe.[28–30] The exploitation of soil has led, and is leading to, widespread landscape degradation, the loss of ecosystem services and the deterioration of the livelihoods of over 250 million people that depend on soils as a source of income.[28,31,32]

3. Causes of global soil degradation are complex and never singular

Soil degradation can occur mainly at three levels; physical, chemical, and biological, with each of these capable of bringing down the ecological value and functionality of landscapes.[1] The extent and intensity of soil degradation varies spatially and temporally making identification and treatment complex (Fig. 1). The drivers that lead to these levels of degradation can be categorized as natural and/or anthropogenic.[33–35] These drivers are

Fig. 1 Soil degradation is a global phenomenon that occurs at different scales, both temporally and spatially. (A) Soil degradation by land abandonment and grazing in south-east Spain (Amoladeras, Almeria). (B) Water and wind erosion in south-east Australia (The Twelve Apostles; Victoria, Australia). (C) Soil and land disturbance in large iron-ore mining in the Pilbara region of north-west Western Australia. (D) Soil sheet and rill erosion processes in south-east Spain (Tabernas, Almeria). (Photo credit: Miriam Muñoz-Rojas and Amber Bateman).

mutually inclusive and interdependent factors that are underlined by social, economic and political factors which contribute to soil degradation and can inhibit restoration efforts (Fig. 2).[36] For example, natural ecosystems cleared for alternative uses, such as agriculture, will suffer from soil degradation through the removal of vegetation and the alteration of natural soil processes. However, land clearing can also amplify natural soil degrading processes such as wind and water erosion. The loss of vegetation can also create salinity issues in the landscape through the rise of the belowground water table. Thus, the causes of land and soil degradation are difficult to identify and quantify with a combination of factors contributing to the degradation of a landscape.[21,37]

3.1 Soil erosion and salinization

Soil erosion is the dominant natural driver of soil degradation worldwide.[38,39] This soil degradation process causes the breakdown of soil

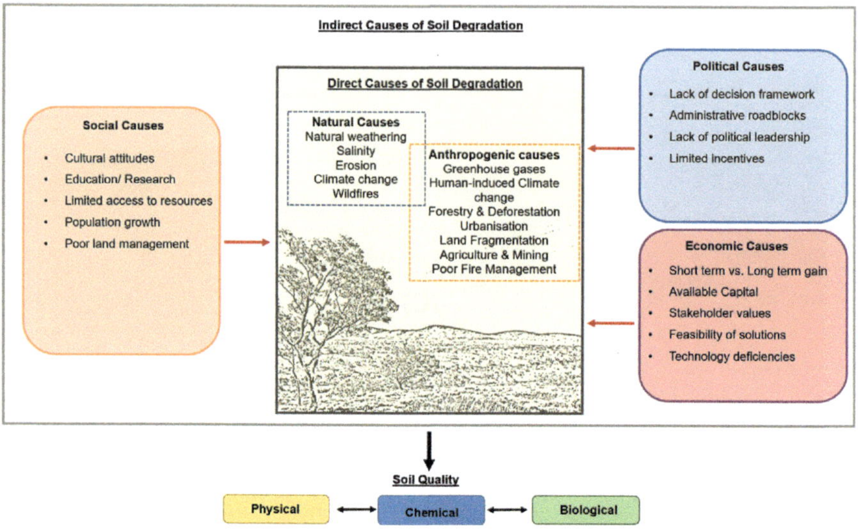

Fig. 2 Direct (natural and anthropogenic) and indirect (political, economic and social) causes of soil degradation. These causes often overlap and can occur simultaneously on a landscape.

aggregates and soil structure through natural weathering.[40–43] In addition, soil erosion amplifies the effects of climate change through releasing minerals into water sources and carbon into the atmosphere.[44] Soil erosion can be caused through weathering processes but also through natural disasters such as floods, landslides and earthquakes.[45,46] At the start of the 21st Century, the global loss of soil as a result of erosion was estimated to be between 24 and 75 billion tonnes.[3,47] In Europe, soil erosion costs between 700 and 14,000 million Euros each year.[48,49] Specifically, water erosion covers 1.1 billion ha and 56% of the world's agricultural land, which has seen declines in productivity.[50] Soil erosion is amplified through land use changes and land clearing for agriculture and is expected to increase due to changes in temperature and rainfall patterns.[50] This damaging process can lead to further land degradation in the form of desertification due to a loss of physical soil properties, soil nutrients and soil biology.[23]

Soil salinization, often the by-product of soil erosion, is a naturally and anthropogenically derived form of soil degradation that can be hazardous to soil quality and have detrimental effects on native plant communities and crop productivity.[34,51] Naturally occuring soil salinity can be caused by poor soil drainage or shallow underground water tables that holds minerals and salts close to the soil surface.[34] Anthropogenically derived soil salinization is predominantly found on agricultural lands with approximately one billion hectares of land affected by salinity world wide.[3,22,52] Causes for soil salinity include the use of fertilizers and the irrigation of land with saline water.[35] As such, soil salinity is a large cause for soil degradation in South European countries with more than 35,000 ha of land affected in Bulgaria.[34]

3.2 Intensive agricultural and land clearing processes

Anthropogenic disturbance of soil goes back almost 13,000 years and today, humans are the largest influencers and movers of soil.[6] Human influenced soil disturbance is now commonly regarded as the sixth soil forming factor.[21] In the context of global change, the additional pressure imposed on soils through anthropogenic influences is of global concern. Anthropogenic disturbances include intensive mining and agricultural practices, inappropriate fire management, urban development, deforestation and landscape fragmentation, compaction, contamination and nutrient imbalance, among others.[21]

Land clearing accounts for approximately 30% of the greenhouse gas emissions while agriculture contributes 10–12% of total anthropogenic greenhouse gas emissions.[53–55] In addition, the use of land cleared for agriculture increases the rate of soil degradation, especially in dryland regions.[23,56] Globally, 38% of the Earth's land surface is covered with agricultural lands.[6] Between the year 2000 and 2012, 2.3 million km^2 of forests were cleared to make way for agricultural expansion.[6,57] In Africa, 55% of land is considered unsuitable for agricultural use yet supports grazing with roughly 30% of the African population dependent on these low fertile soils.[58]

Agricultural and mining practices supports the livelihoods of 40% of the world's population and predominately occurs in drylands, with 45% of the agricultural industry feeding 60% of world population.[6,28,31] However, due to historical weathering through wind and water erosion and poor land management, these landscapes have become degraded and unsuitable for these intensive practices and rehabiliation is extremely challenging.[59] Agricultural practices such as tillage, the use of heavy machinery causing soil compaction, excessive inorganic fertilizers, poor irrigation and water management, and pesticide overuse increase the rate and extent of soil degradation.[48,60–65] These intensive practices directly impact soil quality and ecosystem services such as carbon sequestration and nutrient water cycling which lead to the decline in biodiversity and global health.[63] They also decrease soil resilience to global climate change with an estimated

40% of European agricultural lands vulnerable to moderate or severe levels of soil degradation.[43,66]

Although urban soils fulfill many functions within urban areas, such as supporting buildings and infrastructure, waste management, water purification, pollution retention, inner city green spaces, community gardens and botanic gardens, they are undervalued.[67] Through urbanization, soil quality and soil conservation is seldom considered in city planning beyond their compatability to support infrastructure despite the fact that the management of urban soils can improve the sustainability of cities.[67] This is a growing issue in a world where future projections expect up to 65% of the population to reside in urban areas by 2050.[67–69] Urbanization contributes to soil degradation processess such as water and wind erosion, salinization, desertification, decline in organic matter and soil microbiological activity, and flooding due to land clearing.[70–72] Soil compaction and sealing are processes usually seen on agricultural and urban areas such as industrial and retail sites and transportation infrastructure. In terms of soil sealing, the European commission has published a report assessing effective methods of limiting, mitigating and compensating for the negative effects of soil sealing such as spatial planning for sustainable land use.

4. Consequences of soil degradation are multifaceted

The loss of soil quality can result in both direct and indirect environmental, social and economic consequences (Fig. 3).[36] These consequences occur at the global, national and local level affecting multiple facets of society. Other impacts resulting from the degradation of soils include the loss of cultural identity as well as geological and archeological heritage.[73]

4.1 Food security and freshwater scarcity

Despite agricultural production increasing worldwide, one billion people are still undernourished.[55] With an extra 2.4 billion people to feed by the year 2050, pressure to enhanced crop and livestock production has never been greater.[74,75] Agricultural production will need to intensify by 70%–100% on already cultivated lands to feed the growing population. This agricultural intensification on already cultivated lands is required if we are to reduce land degradation caused by land clearing.[48,76–78] However, over recent years, the productivity of farming land has deteriorated as a result of soil degradation.

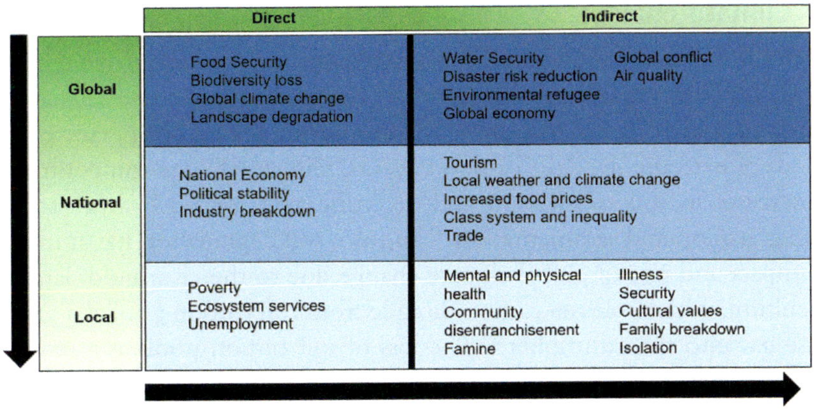

Fig. 3 The direct and indirect consequences of soil and land degradation at the global, national and local scale. *Adapted from Turner KG, Anderson S, Gonzales-Chang, M., et al. A review of methods, data, and models to assess changes in the value of ecosystem services from land degradation and restoration.* Ecol Model. *2016;* ***319****, 190–207.*

In terms of land productivity, it is estimated that global land yield is constantly declining with trends of landscape scale stress and quality losses experienced in 20% of agricultural lands.[79] This is experienced predominantly in drylands across Africa, South America and Australia. Case studies into the agricultural industries of other ecosystem types in China, Ethiopia, Mexico, Uganda and Rwanda, Chile, and Indonesia estimate a continuing decline in land productivity of 3%–7% each year as a result of land and soil degradation.[4,80]

Soils play a large role in the storing and filtering of freshwater through the landscape but as the global population rises, the demand for fresh water increases.[81] However, the degradation of soils though land use changes can impact water quality by altering soil water relations (e.g. water infiltration, retention and evapotranspiration).[81,82] Over large areas, land use changes and soil degradation can alter local rainfall amounts and pattern, ultimately changing water recharge of local sources for fresh water. In addition, the extraction of water from fresh water systems, the excessive use of fertilizers and pesticides on agricultural land, and the discharge of industry waste into river systems can diminish the quality of water for users downstream.[82] The ramifications of neglecting soil conservation and the development of sustainable farming practices will have a direct impact on the health of the worlds poorest people and farmers whom rely on their lands as a source of income.[46,83]

4.2 Climate change

Although the consequences of climate change are somewhat contentious and vary in impact around the globe, it is evident that climate change has contributed to soil degradation.[43,65] Through increased temperatures and periods of extreme drought or flood events, climate change can contribute to increases in soil salinization and desertification, or wash away topsoil causing erosion and sedimentation.[61] In turn, soil degradation has amplified the impact and rate of global climate change due to the clearing of land for agriculture, which releases carbon dioxide from soil carbon pool as a greenhouse gas into the atmosphere. The loss of soil carbon pools as a result of anthropogenic activities is a large contributor to global greenhouse gas emissions. Global soil carbon pools contain more carbon than both the atmospheric and aboveground carbon pools, with 33% of soil carbon stored in the tropics.[84,85] The degradation of soil in systems such as tropical rainforests, native grasslands and coastal wetlands through land clearing for agriculture and urbanization significantly impacts the soil carbon pool.[86] In tropical forests, deforestation accounts for 20% of the anthropogenic carbon dioxide emissions.[56,87] Another example of soil carbon loss is in native grasslands which hold 10% of the globes soil carbon.[26,30] This negative cycle of soil degradation and global climate change increases the vulnerability of soils, agricultural lands and natural ecosystems to global climate change and natural disasters.[1,65] Soil rehabilitation and conservation is crucial for combatting climate change as healthy soils are resilient and will mitigate biodiversity responses to changing rainfall and temperature.[43]

4.3 Social and economic impacts

In addition to the direct environmental and agricultural ramifications of soil degradation, there are a number of indirect social and economic implications for the loss of global soil health.[36] For example, the decline in agricultural productivity as a result of soil degradation in drylands is estimated to be costing developing countries 4%–8% of their national domestic product, annually.[6,88] In several Latin American countries this loss ranges between 8% and 14%.[6,88] Projections estimate that future declines in agricultural productivity could result in a decline in gross domestic product (GDP) of 4% in India, 20% in Bukina Faso and 30% in Mali.[88] This decline of national GDP can have consequences for the country's economy and a government's ability to deliver goods and services to support farmers. In addition, a decline in farmer income reduces investment into sustainable land management. In turn, these consequences create a negative feedback loop where populations that rely on degraded lands and declining farmer income turn into causes of land and soil degradation though limited inputs into soil recovery.[36,89] With time, the soil continues to degrade and the agricultural output of the land continues to decline, increasing local food prices. These environmental and economic problems can lead to social problems such as large scale poverty, famine and illness, a decline in mental and physical health, loss of family and cultural values, security and community disenfranchisement.[4]

5. Future perspectives; possible solutions: to whom does the burden of soil restoration fall?

The direct cost of land degradation to the global economy is estimated to range from 18 billion to 20 trillion USD, annually.[6,90] The cost of losing vital ecosystem services is estimated between 6.3 and 10.6 trillion USD, annually.[6,90] As such, the cost of not taking action against land and soil degradation far outweighs the cost of supporting the development of sustainable management and rehabilitation practices for degraded and non-degraded land.[19]

This increasing threat to the survival of modern society in the face of global climate change and land degradation has driven the global community to take steps in creating international targets for addressing these high impact global issues. In the "2030 Agenda for Sustainable Development" land and soil degradation was recognized by the United Nations General Assembly.[78,91] This agenda initiated 17 SDGs that aim to streamline and unite landholders globally to one universal goal to achieve Zero Net Land Degradation by ceasing land degradation of natural ecosystems and restoring already degraded landscapes. In the context of soil recovery, these global SDGs work toward shaping and driving the methods by which soil undergoes rehabiliation, conservation and sustainable management and the protocols we follow to monitor these management strategies.[35] Since then, the UN General Assembly announced 2021–30 as the decade of ecosystem restoration as a call to action to scale up aggressive restoration efforts.[92] Other soil restoration initiatives include the Bonn challenge which seeks to restore 150 ha by 2020 and 350 ha by 2030.[91,93] First launched in 2011, this initiative seeks to restore degraded forests through afforestation, forest regeneration and agroforestry expansion.[94]

5.1 Farmer perceptions and engagement

Incentive, access to information and resource availability are the dominant hurdles for farmers faced with landscape scale restoration. Research into farmer perspectives indicate that a large number of farmers see the problem of land degradation and climate change as too big an issue for them to tackle individually and thus did not have any plans in place for changing land management strategies.[10,53] Any decision to improve farming methods are strongly dictated by a host of social, economic and environmental factors.[11,53,83,95] Barnes et al. 2012, found that farmers are concerned with the soil quality of their land but also with the debt associated with adopting advanced agricultural practices that preserve soil health.[53] These economic stressors associated with cost of improved farming methods supress voluntary uptake of innovative land management strategies and highlight the need for scientists to engage local farmers and land holders in research and knowledge sharing. In addition, a lot of new technologies developed for use in the agricultural industry focus on the improving productivity in developed countries that can afford the new technology.[96] It is hypothesized that small landholder and farmers, especially those located in developing nations, are isolated from new technologies and land management initiatives.[11,96] Not only do these technologies prove to be ineffectual in the climate and harsh environment of dryland developing countries but this amplifies the social and economical inequalities prevalent in the global sphere. It is crucial that governments work to facilitate access to conservation practices and soil data through educational programs, consultants, national landcare programs, such as Natural Resource Management groups (NRMs), and access to products though subsidies. Currently, the investment by policy-makers into restoration efforts, farming incentives and research falls short of what is necessary to improve soil quality despite a large portion of the wealth of many countries being found predominantly below ground. Nations with large agricultural industries have an opportunity and a responsibility to lead the way in researching, developing and initiating practices to protect and recover soils to secure the future of the countries wealth.

Although the SDGs unify the globe behind a set of global issues they are isolated from farmers and landholders.[7] In their current form, these goals are aimed at world leaders, politicians and scientists but isolate landholders and the general public, the very people who will be impacted the most by these global issues and are ultimately the ones responsible for the on-ground recovery of these degraded landscapes. These 17 goals need to be broken down and options for attaining these goals must be provided. Goals that can be achieved at a local, regional and global level involve the active participation of all factions of society in taking ownership and responsibility for the state of the globe. For example: *Goal 15: Protect, restore and promote sustainable use of terrestrial ecosystems, sustainable manage forests, combat desertification, and halt and reverse land degradation and halt biodiversity loss* (Fig. 4).[3,7,17] This goal has nine sub-goals one of which addresses soil degradation; *15.3. By 2030, combat desertification, restore degraded land and soil, including land affected by desertification, drought and floods, and strive to achieve a land degradation neutral world.* This ambiguous goal, lacks measureable targets of how much land needs to be recovered and its measure of success is only the portion of land that is degraded over the total land area (Fig. 4).[7,8] In addition, they fail to identify the key stakeholders that will be held accountable for their role in ensuring that the SDGs are put into action. The lack of direction offered in these goals opens them up to a broad scope of interpretation which can hold landowners hostage to the discretion of world leaders to pick and choose what suits their agenda. A classification system for identifying degraded land is required and solutions for remediating degraded soil once it has been identified. The development of achievable on-ground targets for soil

Fig. 4 A breakdown of goal 15.3 of the UN "Sustainable Development Goals" (https://sustainabledevelopment.un.org/sdg15) and how it relates to soil science and conservation. Soil chemistry indicators include: pH, electrical conductivity (EC), organic carbon (OrgC), total nitrogen (TN) and phosphorus (P).

conservation, an understanding of the methods to achieve those goal and the means for assessing progress through soil quality indicators are required if the SDGs are to be realized (Fig. 4).

5.2 Soil quality indicators

Soil quality is linked to the health of 80% of ecosystem services and thus should be at the forefront of global restoration focus and outcomes. Using soil quality indicators to assess soil function identifies soil degradation, the causes of degradation and is a tool for monitoring the recovery of soil quality overtime.[97–100] Soil quality indicators include soil chemical indicators such as soil pH and electrical conductivity, organic carbon and total nitrogen, biological indicators such as microbial abundance and diversity and, physical indicators such as aggregate stability, soil texture and structure and porosity.[97,101] All of these indicators come together to create an understanding of the soil status and the overall health of the ecosystem.[97,102–104] Although soil degradation is recognized in the SDGs (2, 3, 11, 13, 14 and 15) as a factor affecting land desertification, there are no assigned indicators of soil quality to evaluate progress in reaching these goals.[101] A range of soil physical, chemical and biological indicators are crucial for understanding soil function across various landscapes and ecosystem types. In turn, these soil functions shed light on the state of land and soil degradation and the capability of land to deliver economic and agricultural goods as well as ecosystem services.[100] Soil functions that can be instrumental in determining land capabilities include nutrient cycling, water and climate regulation, land structure and stability, and soil microbial and plant diversity.[100]

Land capability assessments designed to look at the ability of a piece of land to produce a successful crop should also assess the social and environmental and economic costs of land use change.[55] Soils provide a number of services, be it food, materials or minerals, but the management of these unique landscapes must consider the different characteristics, dynamics and thresholds in a way that utilizes the systems strengths to maximize sustainable output.[73,91,105,106] The use of this strategic spatial planning has large capabilities in contributing to the neutralization of soil degradation. For example, agroforestry is one of many options for improving crop yield while simultaneously conserving ecological functions of the landscape.[21] This method also increases resilience to climate change. Other remedial practices include reforestation, grassland enclosures, windbreak networks; all practices adopted by Russia and China after experiencing intense soil degradation as a result of wind and water erosion due to land clearing.[21]

5.3 Current research and the role of soil scientists

In dryland regions, up to 25% of the terrestrial soils are degraded with these lands classified as the most under threat from global climate change.[65] Strongly relied upon by the local community for a source of food and income, it is crucial that the broader academic, industrial and political community identify the hurdles preventing the sustainable and productive management of lands and design policy that is accessible to small scale farms to assist in securing their future. Soil science has the potential to play a large role in the challenge of dealing with issues such as soil degradation, land productivity and global food security.[48,107] Despite this fact, in the past soil science has received little attention from policy-makers even though soil degradation contributes significantly to the impacts of global climate change.[43] However, with the International Year of Soils declared by the United Nations in 2015 and through increasing global awareness of the importance of soil to provide ecological and economic services, soil science has obtained the global spotlight as a means for landscape restoration.[93] The United Nation's SDGs provide an opportunity for soil scientists to promote the field and demonstrate the value and versatility of soil science globally in shaping future land restoration.[91]

Soil scientists have a large responsibility to increase the accessibility of their research to landholders, farmers, businesses and the general public. There is currently a considerable amount of research looking into remedial practices to increase soil recovery using amendments, mulches, soil microbes and reinstating native vegetation.[76,102,103,108] However, there is a need to bridge the gap with stakeholders and industry partners to up-scale research in a feasible manner.[109] Furthermore, the promotion of interdisciplinary research projects looking at long-term, whole system approaches that collaborate with landholders and farmers will drive the adoption of sustainable and regenerative land management practices.[110,111] Scientific conferences should seek to promote the accessbility of science through open days that invite land managers, farmers, the local indigenous community and the general public to engage in current research. In addition, monetary support for land managers and farmers to fund attendance and workshops should be available to increase accessibility of research. Farmer discussion groups offer another local-scale opportunity for policy-makers and scientists to network and collaborate with groups of end-users (farmers and landholders) to develop research and sustainable land management practices that are designed to address soil and land-based issues in the local region.[53,112]

Furthermore, scientists and policy makers must engage with local indigenous communities to enrich local learning and knowledge from these communities and promote cultural connections with the land.[118]

5.4 The role of government

Soil fertility and the level of soil degradation of a region is a strong indicator for the productivity of the land.[15,97] However, this is a realm where social and economical drivers are prevalent and often are prioritize over the environment and social well-being of the local community. Soil is often overlooked in national and international policy-making when evaluating vital resources that contribute to ecosystem services and sustainability. An example of this is in the classification of ecosystem types for economical evaluation in *The Economics of Ecosystems and Biodiversity* (TEEB).[113] The TEEB is a global initiative designed to raise awareness and support policy-makers in acknowledging the importance of biodiversity and ecosystem services. However, the conservation of soils was neglected in the framework of the program. Policy-makers often include the conservation of soils as a by-product of other policy objectives and soils are not considered as an independent entity.[114] In addition, the investment by policy-makers into restoration efforts, farming incentives and research is limited. A clearer understanding of the implications of soil degradation to country wealth is required to highlight the importance of soil quality and its conservation and governments need to offer opportunities and incentives for the collaboration of scientists to work with farmers and industry.[55,115] If governments do not invest in the restoration of soils globally, we will see an increase in environmental refugees leaving degraded regions as a result of low agricultural production and poverty.

Limited access to quality education, soil knowledge and informative resources give rise to poor land management, global social inequality and poverty.[11,83,116,117] As such, the promotion of soil science, soil degradation and conservation within the education system is crucial for raising awareness of the importance of soils.[106] Governments and soil scientist must take an active role to improve the transfer of knowledge for not only end-users and policy-makers to the general public through the education system.[91]

6. Conclusion

Under current and projected global environmental change, the Earth's soils are undergoing damaging and irreversible levels of degradation. The implication of soil degradation extends beyond only environmental concerns and reaches into society's social, economic and cultural framework. However, the causes for soil degradation are multifaceted with landscapes afflicted by both natural and anthropogenic factors that require coordinated and collaborative efforts to overcome them. To create a sustainable future and combat global threats to humankind, the United Nations General Assembly defined 17 Sustainable Development Goals. One of these goals sets out to reverse land and soil degradation by striving for a land degradation neutral world. Although these goals identify crucial issues and establish global targets for addressing these issues, the goals are isolated from farmers and landholders that are ultimately going to carry the responsibility of conserving our soils and land. There is limited support or guidance as to how to achieve these SDGs, the methods for monitoring progress or the desired target. As the impacts of soil degradation are and will be far reaching and effect all members of society, all of society must support the adaptation and establishment of sustainable land management practices by increasing accessbility of innovative technology, science and knowledge to farmers and landholders. Through stakeholder collaborations, science can be upscaled to deliver high impact research outputs that inform management practices. Furthermore, the integration of soil science into the education system and creating government policy that supports and creates incentives for the adoption of sustainable land management to assist in on-ground action is required. For at the end of the day, when it comes to soil degradation, climate change and food security, we are all stakeholders.

References

1. Lal R. Restoring soil quality to mitigate soil degradation. *Sustainability* 2015;**7**:5875–95.
2. *Department of economic and social affairs world population projected to reach 9.7 billion by 2050.* New York: United Nations; 2015.
3. Stavi I, Lal R. Achieving zero net land degradation: challenges and opportunities. *J Arid Environ* 2015;**112**:44–51.
4. Barbier EB, Hochard JP. Does land degradation increase poverty in developing countries? *PLoS One* 2016;**11**:e0152973.
5. Chasek P, Akhtar-Schuster M, Orr BJ, Luise A, Ratsimba HR, Safriel U. Land degradation neutrality: the science-policy interface from the UNCCD to national implementation. *Environ Sci Policy* 2019;**92**:182–90.
6. United Nations Convention to Combat Desertification. *The global land outlook.* Bonn, Germany. 2017.
7. Liverman DM. Geographic perspectives on development goals: constructive engagements and critical perspectives on the MDGs and the SDGs. *Dialog Human Geograph* 2018;**8**:168–85.
8. Moseley WG. Geography and engagement with UN development goals: rethinking development or perpetuating the status quo? *Dialog Human Geograph* 2018;**8**:201–5.

9. Eckermann L. *Health promotion principles as a catalyst for translating the SDGs into more transformative action*. Oxford University Press; 2016.
10. Tesfahunegn GB, Mekonen K, Tekle A. Farmers' perception on causes, indicators and determinants of climate change in northern Ethiopia: implication for developing adaptation strategies. *Appl Geogr* 2016;**73**:1–12.
11. Deressa TT, Hassan RM, Ringler C, Alemu T, Yesuf M. Determinants of farmers' choice of adaptation methods to climate change in the Nile Basin of Ethiopia. *Glob Environ Chang* 2009;**19**:248–55.
12. Lal R. Soil degradation by erosion. *Land Degrad Dev* 2001;**12**:519–39.
13. Robinson DA, Emmett BA, Reynolds B, et al. Soil natural capital and ecosystem service delivery in a world of global soil change. *Soils Food secur* 2012;**30**:41.
14. Brevik E, Cerdà A, Mataix-Solera J, et al. The interdisciplinary nature of SOIL. *Soil* 2015;**1**:117–29.
15. Pereira P, Bogunovic I, Muñoz-Rojas M, Brevik EC. Soil ecosystem services, sustainability, valuation and management. *Curr Opin Environ Sci Health* 2017;**5**:7–13.
16. Abrahams P. Soils: their implications to human health. *Sci Total Environ* 2002;**291**:1–32.
17. United Nations Convention to Combat Desertification. *Policy brief: zero net land degradation*. Bonn, Germany. 2012.
18. Lal R, Safriel U, Boer B. *Zero net land degradation: a new sustainable development goal for Rio+ 20. Proccedings of Conference Zero net land degradation: A new sustainable development goal for Rio+*, vol. 20; 2012.
19. Nkonya E, Mirzabaev A, Von Braun J. *Economics of land degradation and improvement: a global assessment for sustainable development*. Springer; 2016.
20. Bai ZG, Dent DL, Olsson L, Schaepman ME. Proxy global assessment of land degradation. *Soil Use Manage* 2008;**24**:223–34.
21. Karlen DL, Rice CW. Soil degradation: will humankind ever learn? *Sustainability* 2015; **7**:12490–501.
22. Gibbs H, Salmon JM. Mapping the world's degraded lands. *Appl Geogr* 2015;**57**:12–21.
23. Hernández-Becerra N, Tapia-Torres Y, Beltrán-Paz O, Blaz J, Souza V, García-Oliva F. Agricultural land-use change in a Mexican oligotrophic desert depletes ecosystem stability. *PeerJ* 2016;**4**:e2365.
24. Worrall R, Neil D, Brereton D, Mulligan D. Towards a sustainability criteria and indicators framework for legacy mine land. *J Clean Prod* 2009;**17**:1426–34.
25. Unger C, Lechner A, Glenn V, Edraki M, Mulligan D. Mapping and prioritising rehabilitation of abandoned mines in Australia. *Proceedings Life-of-Mine* 2012:259–66.
26. Leite LFC, Sagrilo E, Galvao SRS, Maciel GA. *Carbon sequestration in no-tillage systems on tropcial soils of Brazil*. 2010.
27. Bisaro A, Kirk M, Zdruli P, Zimmermann W. Global drivers setting desertification research priorities: insights from a stakeholder consultation forum. *Land Degrad Dev* 2014;**25**:5–16.
28. Wang T, Xue X, Zhou L, Guo J. Combating aeolian desertification in northern China. *Land Degrad Dev* 2015;**26**:118–32.
29. Millennium Ecosystem Assessment. *Ecosystems and human wellbeing: desertification synthesis*. Washington, DC. 2005.
30. Abdalla K, Mutema M, Chivenge P, Everson C, Chaplot V. Grassland degradation significantly enhances soil CO_2 emission. *Catena* 2018;**167**:284–92.
31. Reynolds J, Stafford Smith D, Lambin E, et al. Global desertification: building a science for dryland development. *Science* 2007;**316**:847–51.
32. Gonzalez-Roglich M, Zvoleff A, Noon M, et al. Synergizing global tools to monitor progress towards land degradation neutrality: trends. Earth and the World Overview of Conservation Approaches and Technologies sustainable land management database. *Environ Sci Policy* 2019;**93**:34–42.

33. Karlen DL, Rice CW. *Soil degradation: will humankind ever learn?* Multidisciplinary Digital Publishing Institute; 2015.
34. Penov I, Manolov I, Alexiev A, Kavardziev Y. Salinisation in Bulgaria: institutional settings for soil conservation (a case study of Belozem village). *Land Degrad Dev* 2011;**22**:134–43.
35. Tóth G, Montanarella L, Rusco E. *Threats to soil quality in Europe*. 2008.
36. Turner KG, Anderson S, Gonzales-Chang M, et al. A review of methods, data, and models to assess changes in the value of ecosystem services from land degradation and restoration. *Ecol Model* 2016;**319**:190–207.
37. Anjum SA, Wang LC, Xue LL, Saleem MF, Wang GX, Zou CM. Desertification in Pakistan: causes, impacts and management. *J Food Agric Environ* 2010;**8**:1203–8.
38. Panagos P, Borrelli P, Meusburger K, et al. Global rainfall erosivity assessment based on high-temporal resolution rainfall records. *Sci Rep* 2017;**7**.
39. Oldeman LR. Global extent of soil degradation. In: *Bi-annual report 1991–1992/ ISRIC*. ISRIC; 1992. p. 19–36.
40. Kostecka J, Mazur-Pączka A, Pączka G, Garczyńska M. Lumbricidae biodiversity at the sites in Bieszczady Mountains (Poland) after 25 years. *J Ecolog Eng* 2018;**19**:2.
41. Rossati A. Global warming and its health impact. *Int J Occup Environ Med* 2017;**8**. 963-967-920.
42. Giertz S, Hiepe C, Hollermann B, Diekkruger B. *Impacts of global change on water resources and soil degradation in Benin*. 2010.
43. Balkovic J, Skalsky R, Folberth C, et al. Impacts and uncertainties of+2 degrees C of climate change and soil degradation on European crop calorie supply. *Earths Future* 2018;**6**:373–95.
44. Lal R. Soil erosion and the global carbon budget. *Environ Int* 2003;**29**:437–50.
45. Masyagina O, Evgrafova SY, Bugaenko T, et al. Permafrost landslides promote soil CO_2 emission and hinder C accumulation. *Sci Total Environ* 2019;**657**:351–64.
46. Brausmann A, Bretschger L. Economic development on a finite planet with stochastic soil degradation. *Eur Econ Rev* 2018;**108**:1–19.
47. Secretariat U. Land and soil in the context of a green economy for sustainable development, food security and poverty eradication, the submission of the UNCCD secretariat to the preparatory process for the Rio+ 20 Conference. *Revised Version* 2011;**18**.
48. Bindraban PS, van der Velde M, Ye L, et al. Assessing the impact of soil degradation on food production. *Curr Opin Environ Sustain* 2012;**4**:478–88.
49. Montanarella L. Trends in land degradation in Europe. In: *Climate and land degradation*. Springer; 2007. p. 83–104.
50. Correa SW, Mello CR, Chou SC, Curi N, Norton LD. Soil erosion risk associated with climate change at Mantaro River basin, Peruvian Andes. *Catena* 2016;**147**: 110–24.
51. Škarková P, Kos M, Drobne D, Vávrová M, Jemec A. Effects of food salinization on terrestrial crustaceans *Porcellio scaber*. *Appl Soil Ecol* 2016;**100**:1–7.
52. Squires VR, Glenn EP. *Salination, desertification, and soil erosion. The role of food, agriculture, forestry and fisheries in human nutrition*, vol. 3; 2011. p. 102–23.
53. Barnes AP, Toma L. A typology of dairy farmer perceptions towards climate change. *Clim Change* 2012;**112**:507–22.
54. Metz B. *Climate change 2007: mitigation: contribution of working group III to the fourth assessment report of the intergovernmental panel on climate change*. Intergovernmental Panel on Climate Change; 2007.
55. Sachs J, Remans R, Smukler S, et al. Monitoring the world's agriculture. *Nature* 2010; **466**:558.

56. Tesfaye MA, Bravo F, Ruiz-Peinado R, Pando V, Bravo-Oviedo A. Impact of changes in land use, species and elevation on soil organic carbon and total nitrogen in Ethiopian Central Highlands. *Geoderma* 2016;**261**:70–9.
57. Hansen MC, Potapov PV, Moore R, et al. High-resolution global maps of 21st-century forest cover change. *Science* 2013;**342**:850–3.
58. Bationo A, Hartemink A, Lungo O, et al. *African soils: their productivity and profitability of fertilizer use: background paper for the African Fertilizer Summit 9–13th June 2006.* Abuja, Nigeria. 2006.
59. Anaya-Romero M, Abd-Elmabod SK, Muñoz-Rojas M, et al. Evaluating soil threats under climate change scenarios in the andalusia region, southern Spain. *Land Degrad Dev* 2015;**26**:441–9.
60. Van Oost K, Govers G, De Alba S, Quine T. Tillage erosion: a review of controlling factors and implications for soil quality. *Prog Phys Geogr* 2006;**30**:443–66.
61. Misra AK. Climate change impact, mitigation and adaptation strategies for agricultural and water resources, in Ganga Plain (India). *Mitig Adapt Strategies Glob Change* 2013;**18**: 673–89.
62. Krummelbein J, Peth S, Horn R. Determination of pre-compression stress of a variously grazed steppe soil under static and cyclic loading. *Soil Tillage Res* 2008;**99**: 139–48.
63. Oliveira E, Tobias S, Hersperger AM. Can strategic spatial planning contribute to land degradation reduction in urban regions? State of the art and future research. *Sustainability* 2018;**10**:949.
64. Eftene CA, Dumitru S, Manea A, Răducu D. *A review of the impacts of soil sealing on soil properties in Romania. International Multidisciplinary Scientific GeoConference: SGEM: surveying geology & mining ecology management*, **2**; 2016. p. 9–13.
65. Thornton PK, Ericksen PJ, Herrero M, Challinor AJ. Climate variability and vulnerability to climate change: a review. *Glob Chang Biol* 2014;**20**:3313–28.
66. Lal R. Soil carbon stocks under present and future climate with specific reference to European ecoregions. *Nutrient Cycl Agroecosyst* 2008;**81**:113–27.
67. Morel JL, Chenu C, Lorenz K. Ecosystem services provided by soils of urban, industrial, traffic, mining, and military areas (SUITMAs). *J Soils Sediments* 2015;**15**:1659–66.
68. Cohen JE. Human population: the next half century. *Science* 2003;**302**:1172–5.
69. United Nations Department of Economic and Social Affairs Population Division. *World urbanization prospects: the 2014 revision.* 2015.
70. Bajocco S, Salvati L, Ricotta C. Land degradation versus fire: a spiral process? *Prog Phys Geogr* 2011;**35**:3–18.
71. Stolte J, Tesfai M, Oygarden L, et al. *Soil threats in Europe: status, methods, drivers and effects on ecosystem services.* 2016.
72. Panagos P, Van Liedekerke M, Jones A, Montanarella L. European soil data centre: response to European policy support and public data requirements. *Land Use Policy* 2012;**29**:329–38.
73. Bouma J, Broll G, Crane TA, et al. Soil information in support of policy making and awareness raising. *Current Opinion in Environmental Sustainability* 2012;**4**:552–8.
74. Singh BK, Trivedi P, Singh S, Macdonald CA, Verma JP. Emerging microbiome technologies for sustainable increase in farm productivity and environmental security. *Microbiol Australia* 2018.
75. Waldron A, Garrity D, Malhi Y, Girardin C, Miller D, Seddon N. Agroforestry can enhance food security while meeting other Sustainable Development Goals. *Trop Conserv Sci.* 2017;**10**. 1940082917720667.
76. Singh BK, Trivedi P, Singh S, Macdonald CA, Verma JP. Emerging microbiome technologies for sustainable increase in farm productivity and environmental security. *Microbiol Australia* 2018;**39**:17–23.

77. Waldron A, Garrity D, Malhi Y, Girardin C, Miller DC, Seddon N. Agroforestry can enhance food security while meeting other sustainable development goals. *Trop Conserv Sci* 2017;**10**.
78. Chasek P, Safriel U, Shikongo S, Fuhrman VF. Operationalizing Zero Net Land Degradation: the next stage in international efforts to combat desertification? *J Arid Environ* 2015;**112**:5–13.
79. Zika M, Erb KH. The global loss of net primary production resulting from human-induced soil degradation in drylands. *Ecol Econ* 2009;**69**:310–8.
80. The Global Commission on the Economy and Climate. *Chapter 3 - land use. Better growth, better climate: the new climate economy report.* 2014.
81. Peters NE, Meybeck M. Water quality degradation effects on freshwater availability: impacts of human activities. *Water Int* 2000;**25**:185–93.
82. Castello L, Macedo MN. Large-scale degradation of Amazonian freshwater ecosystems. *Glob Chang Biol* 2016;**22**:990–1007.
83. Mertz O, Mbow C, Reenberg A, Diouf A. Farmers' perceptions of climate change and agricultural adaptation strategies in rural Sahel. *Environ Manag* 2009;**43**:804–16.
84. Lemma B, Kleja DB, Nilsson I, Olsson M. Soil carbon sequestration under different exotic tree species in the southwestern highlands of Ethiopia. *Geoderma* 2006;**136**: 886–98.
85. Lal R. Soil carbon sequestration to mitigate climate change. *Geoderma* 2004;**123**:1–22.
86. Brevik EC. Soils and climate change: gas fluxes and soil processes. *Soil Horiz* 2012;**53**: 12–23.
87. Baccini A, Laporte N, Goetz S, Sun M, Dong H. A first map of tropical Africa's above-ground biomass derived from satellite imagery. *Environ Res Lett* 2008;**3**:045011.
88. United Nations United Nations Convention to Combat Desertification. Desertification: The invisible frontline; 2013.
89. Mbow C, Mertz O, Diouf A, Rasmussen K, Reenberg A. The history of environmental change and adaptation in eastern Saloum-Senegal-Driving forces and perceptions. *Glob Planet Chang* 2008;**64**:210–21.
90. Costanza R, de Groot R, Sutton P, et al. Changes in the global value of ecosystem services. *Glob Environ Chang* 2014;**26**:152–8.
91. Keesstra SD, Bouma J, Wallinga J, et al. The significance of soils and soil science towards realization of the United Nations Sustainable Development Goals. *Soil* 2016; **2**:111–28.
92. United Nations Environment Programme. *New UN Decade on Ecosystem Restoration offers unparalleled opportunity for job creation, food security and addressing climate change.* 2019.
93. Bouma J, Montanarella L. Facing policy challenges with inter-and transdisciplinary soil research focused on the UN Sustainable Development Goals. *Soil* 2016;**2**:135–45.
94. Laestadius L, Maginnis S, Minnemeyer S, Potapov P, Saint-Laurent C, Sizer N. Opportunities for forest landscape restoration. *Unasylva* 2011;**62**:238.
95. Ervin CA, Ervin DE. Factors affecting the use of soil conservation practices: hypotheses, evidence, and policy implications. *Land Economics* 1982;**58**:277–92.
96. Tittonell P, Giller KE. When yield gaps are poverty traps: the paradigm of ecological intensification in African smallholder agriculture. *Field Crop Res* 2013;**143**:76–90.
97. Costantini EA. Soil indicators to assess the effectiveness of restoration strategies in dryland ecosystems. *Solid Earth* 2016;**7**:397.
98. Pulido M, Schnabel S, Contador JFL, Lozano-Parra J, Gómez-Gutiérrez Á. Selecting indicators for assessing soil quality and degradation in rangelands of Extremadura (SW Spain). *Ecol Indicat* 2017;**74**:49–61.
99. Mukhopadhyay S, Maiti S, Masto R. Development of mine soil quality index (MSQI) for evaluation of reclamation success: a chronosequence study. *Ecol Eng* 2014;**71**:10–20.

100. Muñoz-Rojas M. Soil quality indicators: a critical tool in ecosystem restoration. *Curr Opin Environ Sci Health* 2018;**5**:47–52.
101. Jónsson JÖG, Davíðsdóttir B, Jónsdóttir EM, Kristinsdóttir SM, Ragnarsdóttir KV. Soil indicators for sustainable development: a transdisciplinary approach for indicator development using expert stakeholders. *Agric Ecosyst Environ* 2016;**232**:179–89.
102. Prosdocimi M, Jordán A, Tarolli P, Keesstra S, Novara A, Cerdà A. The immediate effectiveness of barley straw mulch in reducing soil erodibility and surface runoff generation in Mediterranean vineyards. *Sci Total Environ* 2016;**547**:323–30.
103. Benigno SM, Dixon KW, Stevens JC. Increasing soil water retention with native-sourced mulch improves seedling establishment in postmine Mediterranean sandy soils. *Restor Ecol* 2012;**21**:617–26.
104. Jordán A, Zavala LM, Muñoz-Rojas M. Mulching, effects on soil physical properties. In: *Encyclopedia of agrophysics*. Springer; 2011. p. 492–6.
105. Boyle G, Schulte R, Donnellan T, et al. Food harvest 2020: green white gold. In: *Proceedings of conference food harvest 2020*. Green White Gold; 2011.
106. Towers W, Creamer R, Broll G, et al. *Soil awareness and education: developing a pan European approach*. 2010.
107. Wingeyer A, Amado T, Pérez-Bidegain M, et al. Soil quality impacts of current South American agricultural practices. *Sustainability* 2015;**7**:2213–42.
108. Abbott L, Macdonald L, Wong M, Webb M, Jenkins S, Farrell M. Potential roles of biological amendments for profitable grain production—A review. *Agric Ecosyst Environ* 2018;**256**:34–50.
109. Miller BP, Sinclair EA, Menz MH, et al. A framework for the practical science necessary to restore sustainable, resilient, and biodiverse ecosystems. *Restor Ecol* 2017;**25**(4): 605–17.
110. Perring MP, Erickson TE, Brancalion PH. Rocketing restoration: enabling the upscaling of ecological restoration in the Anthropocene. *Restor Ecol* 2018;**26**(6):1017–23.
111. Riva MJ, Daliakopoulos IN, Eckert S, Hodel E, Liniger H. Assessment of land degradation in Mediterranean forests and grazing lands using a landscape unit approach and the normalized difference vegetation index. *Appl Geogr* 2017;**86**:8–21.
112. Yageta Y, Osbahr H, Morimoto Y, Clark J. *Comparing farmers' qualitative evaluation of soil fertility with quantitative soil fertility indicators in Kitui County*. Kenya: Geoderma; 2019.
113. de Groot R, Fisher B, Christie M, et al. *Integrating the ecological and economic dimensions in biodiversity and ecosystem service valuation*. 2010.
114. Scherr SJ. Soil degradation: a threat to developing-country food security by 2020? *Intl Food Policy Res Inst* 1999.
115. Mihalache M, Ilie L, Marin DI. Romanian soil resources - 'Healthy soils for a healthy life'. Scientific Papers, Series *AgroLife J* 2015;**4**:101–10.
116. Pretty J, Toulmin C, Williams S. Sustainable intensification in African agriculture. *Int J Agric Sustain* 2011;**9**:5–24.
117. Tittonell P. Livelihood strategies, resilience and transformability in African agroecosystems. *Agric Syst* 2014;**126**:3–14.
118. One Earth. *Learning from Indigenous Populations and Local Communities*. 2019. https://doi.org/10.1016/ j.oneear.2019.07.002.

CHAPTER TWO

Agriculture and grazing environments

Igor Bogunovic[a,]*, Manuel Pulido Fernández[b], Ivica Kisic[a] and Maria Burguet Marimón[c]
[a]University of Zagreb Faculty of Agriculture, Department of General Agronomy, Zagreb, Croatia
[b]GeoEnvironmental Research Group, Department of Arts and Territory Sciences, University of Extremadura, Cáceres, Spain
[c]Department of Geography, University of Valencia, Valencia, Spain
*Corresponding author: E-mail: igor.bogunovic@gmail.com

Contents

Abstract

The increase in world population increases the pressure on soil resources and triggers land degradation. Agricultural and grazing environments are one of the most vulnerable ecosystems to this process. Degradation affects importantly soil fertility. Millions of hectares are affected and soil productivity is endangered. This decreases soil ability to provide ecosystem services in quantity and quality. Managers and decision-makers should be open to adopt environmentally-friendly practices and contribute to land degradation neutrality. This chapter focuses on the problems induced by some of the most common soil degradation processes caused by in agriculture and grazing activities. Review the main causes of degradation: depletion of organic matter and biological diversity, structural stability, erosion, compaction, crusting, acidification, alkalization and salinization. Each process (subsection) has a description of the negative impacts of agriculture and grazing activities and potential solutions for sustainable management and restoration.

Advances in Chemical Pollution, Environmental Management and Protection, Volume 4
ISSN 2468-9289
https://doi.org/10.1016/bs.apmp.2019.07.005

Keywords: Soil ecosystem services; Conservation practices; Land degradation; Management; Restoration.

1. Introduction

In the beginning of 2019, the world population was estimated to be almost 7.7 billion. In 1950 we were only 2.5 billion.[1] Nowadays we have tree times more population than in 1950 and this increased food demand and the pressure on soil resources.[2] Both aspects of nutrition - food security and food safety — are important issues in decision makers agenda,[3–5]therefore, new forms of agriculture production are needed to reduce the environmental impact and the ecological footprint.[6]

One of the targets important to achieve under United Nations (UNs) Sustainable Development Goals context is *Land degradation neutrality.*[7] Presently, food distribution is extremely unbalanced. In western world there is surplus in food production and waste generation,[8,9] while in other parts population is facing daily hunger and subnutrition.[10] It is estimated that 800 million people are affected by subnutrition.[11,12] Pretty et al.[13] stated that presently in the western world agricultural productions is starting to be focused on food quality, while in the rest of the world, the main concern is food production. High food production is achieved by the application unsustainable agriculture practices (e.g. mineral fertilizer, pesticides and herbicides). As consequence of this, modern agriculture has become the main diffuse source of soil and water pollution.[14,15]

In the last two decades, organic farming and similar techniques have been recognized as a plausible alternative to conventional practices. However, several works defend that organic farming is not enough to meet global food demand.[16] Some authors[16–18] indicate that yields in organic farming are lower compared to conventional agriculture. However, organic farming practices are more environmentally friendly and produce food with high quality. It is clear that the investment in organic farming will return in the future through multiple environmental benefits and climate change mitigation.[19] In this context, organic farming should not be observed only through the achieved yield, but for other benefits as well.

Agricultural intensive practices changed the "face of planet Earth" by clearing a large area occupied by forests, reducing the biodiversity and increasing the degradation of the ecosystems where they are established and off-site (e.g. freshwater and oceans pollution).[20–22] The direct influence of the human activities is so evident that we entered in a new geological period, the Anthropocene.[23] As consequence of the unbearable and irreversible impacts, it is clear that agriculture production cannot be based on conventional agriculture and more sustainable practices are urgently needed.[13]

There is no doubt that the innovations produced by modern science such as biotechnology,[24] computer technology,[25] biological-genetic engineering[26] open unlimited possibilities. Equally, it is clear that these achievements should be used in the improvement of sustainable agriculture practices at a global scale.[27] Globalization made of earth a 'global village', therefore, any deviation from the concept of sustainability brings risks and threats of global dimension.[28,29] "Ecological conscience" is a term that should be adopted with consistence by public and stakeholders at different levels, creating new ethical principles in environmental management, as shown in Fig. 1.

The principles described can be applied at different scales and socio-economical contexts to increase population wellbeing.[30] For this reason, sustainable agriculture is a decisive factor in the survival and economic progress of the different societies and global socio-economic stability.[5]

Grazing is as one of the most ancient forms of interaction between soil, vegetation and humans. Zeder[31] stated the domestication of a wide variety of animals started in the *Neolithic Revolution* (12,000–9,000 years BP,[32]). It is commonly accepted the sedentarism was the main cause attracted wild animals, particularly dogs (*Canus lupus familiaris*), near human settlements

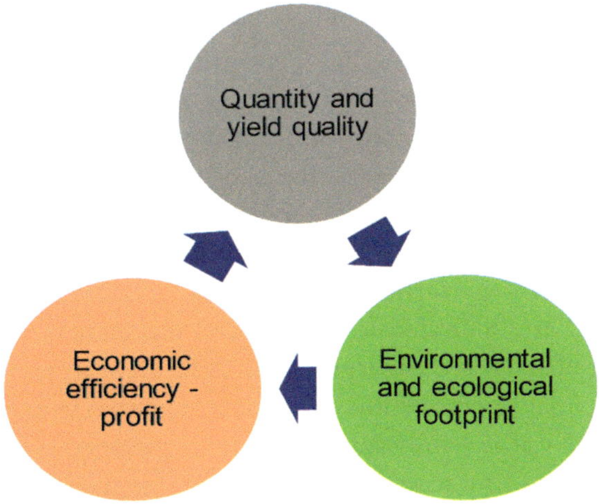

Fig. 1 Relationship between the components of plant breeding and animal husbandry in agriculture.

because the increase of food waste that could be easily consumed.[33] Environmental changes and cold periods such as the Younger Dryas (12,700–11,500 BP) in Europe and droughts in the Middle East forced humans to accelerate the process of cropping and animal domestication due to the scarcity of food.[34]

The first livestock species domesticated by humans were probably the goat (*Capra aegagrus hircus*), 10,000 years ago in the Zagros Mountains (Iran),[35] and sheep (*Ovis orientalis aries*).[36] By 6,000 (B)C. agro-pastoralism already represented the main system of food production in the Northern Hemisphere.[37] In fact, the Bible reports conflicts between agricultural farmers (leaded by Cain) and livestock ranchers (leaded by Abel) in Genesis chapter. Nowadays, 1,500 million pigs, 800 million sheep, 660 million goats and 575 million cattle, amongst other many species, are raised in extensive rangelands, intensive farms or mixed systems.[38]

Soil degradation in grazing environments shall be interpreted as a consequence of past and/or present mismanagement, i.e. situations in which carrying and recovery capacities of a site have been exceeded as consequence of inadequate decisions motivated by several contently-dependent driving forces. Nevertheless, the concept of carrying capacity has been considered as confusing by some authors[39] and its overcoming is simply attributed to a situation of overgrazing,[40] that neglects, for instance, the effects of grazing exclusion,[41] undergrazing[42] or other pastoral practices have improved land conditions keeping or increasing the number of animals due to changes of the paradigm of traditional management (e.g., holistic management,[43] rotational Voisin system,[44] etc.).

On one hand, pastoralism is important in terms of geographical extent and economy as well as cultural legacy (e.g., nomadism cultures, rangeland tenure, etc.). For instance, grazing lands (mainly rangelands and grasslands) cover more than 30% of the global surface and contribute up to 50% of total agricultural production.[45] On the other hand, overgrazing is considered as the main cause of soil degradation in 6.9 million km^2 of land surface, i.e. equivalent to 90% of the territory of Australia.[46] In addition, 470 million ha of arable land are dedicated to animal feed production.[47] Nonetheless, in spite of this global figures soil degradation induced by grazing must be understood as a spatial phenomenon affected by regional patterns of rainfall, local characteristics of soil and topography and by socioeconomic reasons.[48]

The real cause of soil degradation is, indeed, the level of grazing.[49] For instance, too much grazing can lead to soil degradation and loss of biodiversity, while undergrazing decrease grassland areas (conversion into shrubland).

Besides the loss of biodiversity caused by an intensive grazing,[50] negative effects induced by heavy grazing on soil quality (by soil erosion and compaction) and pasture production[51] have been also reported. Controversial results about the role of grazing on soil organic matter (SOM) (i.e., it increases or decreases SOM content) have been found[52,53] and there are other many other degradation processes such as crusting, acidification or leaching, amongst others that should be paid more attention by research community.

The objective of this work is to revise: [1] the role played by agriculture and grazing (management) on soil degradation processes such as depletion of SOM; [2] the lost of biological diversity, structural instability that can lead to soil erosion, compaction and crusting (sealing), and [3] acidification and salinization (alkalinization). Fig. 2 synthesizes the effects of grazing at global scale.

2. Depletion of organic matter and biological diversity

Agricultural intensification in the period between 1950 and 2000 had a strong impact on the homogenization of agricultural landscapes and loss of natural and seminatural habitats. Intensification reduces importantly the

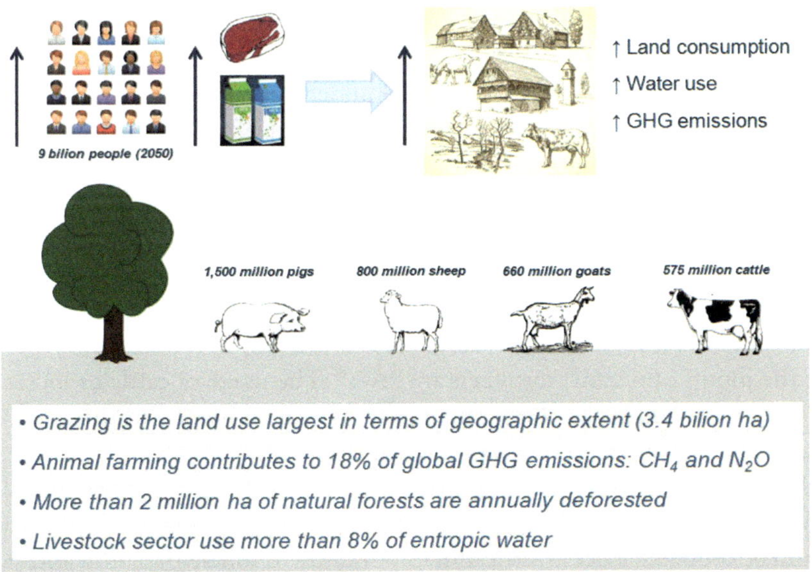

Fig. 2 Scheme about the foreseen number of people in 2050 and livestock animals and the main environmental consequences of grazing reported in the literature so far.

biodiversity. Under title "Our life insurance, our natural capital",[54] the European Union (EU) has developed a strategy for biodiversity conservation by 2020, which includes six basic and mutually dependent goals: [1] to implement Birds and Habitats Directives, [2] to establish green infrastructure and to restore 15% of degraded ecosystems in the EU, [3] to reform the Common Agricultural Policy, [4] to reform the Common Fisheries Policy, [5] to combat invasive alien species and [6] to step up its contribution in the combat to global biodiversity loss. These EU Biodiversity Strategy 2020 targets are in consonance with the UN Sustainable Development Goals (SDGs), thought to achieve a better and more sustainable for all, particularly with the SGD 15 "Life on Land",[55] which requires states to stop biodiversity loss and land degradation.[56] Restoration and protection implies a great number of soil management practices that can reverse land degradation and have positive impact on biodiversity such as the use of permanent or semi-permanent soil cover, minimum tillage and frequent crop rotations.[57]

Degradation process from agricultural activities have important consequences for soil quality such as the decline of SOM, reducing soil structure, water infiltration and increasing compaction.[58] Nowadays, in many areas of the world SOM content in agricultural soils is insufficient for farming. Soils are not productive without the application of fertilisers. The main causes of SOM reduction in agricultural soils are (1) intensive tillage that causes aeration, (2) increase of microbiological activity and (3) humus mineralization.[59] For example, the most important soils in the Pannonian region of Europe have an average of 2.0%–2.5% of organic matter in their arable soil layer, twice lower compared the same soil type under forest cover.[60] The quantity and quality of SOM in agricultural areas is not sufficient for crop production. To meet this demand, farmers apply liquid manure to increase productivity. The operation of spreading the liquid manure is carried out with heavy machinery. The tractors used induce the highest load per unit of area, with a high number of passes on soil outside the vegetation period, in a bare soil highly vulnerable to degradation.[61] At the same time, the number of cattle in all agricultural regions (especially in eastern Europe) reduced and therefore the inputs of organic fertilizers are low.[62] The usage of cattle to increase soil fertility would be less deterimental to the soil compared to the application of liquid manure with heavy machinery.

In order to decrease soil degradation processes, tillage practices and the time and method of fertilizer application have to change.[63,64] Plowing should be reduced, since contributes importantly to soil degradation and carbon dioxide (CO_2) emissions.[65–67] High tillage intensity (mostly plow) reduce SOM due to the high microbial activity[68] and rapid carbon oxidation.[69] Contrary, conservation tillage practices such as no tillage or reduced tillage are very effective in improving SOM, bulk density,[70] porosity, water infiltration,[71] aggregate stability[72] and reduce erosion.[73] However, these benefits may depend on the cropping system, climate and soil type.[74]

In the future, soil tillage should be shallower (by shallow inversing tillage or no-tillage) due to fuel and costs savings, and only occasionally deeper (by loosening). It is realistic then to expect that SOM will increase in the topsoil and reduce CO_2 emissions due to the reduced weight.[75] This management contributes to crop productivity and reduce greenhouse emissions.[76] Soil tillage has to be carried out considering soil type, topography, crop rotation, available machinery and tools, and carbon retention from plant residues in soil.[77] This contributes to the reduction of SOM mineralization.[66] Nevertheless, the most effective way to reduce greenhouse emissions is to adopt reduce or no-tillage management.[75] There are several controversies about the usage of these practices,[78] like higher compaction, poor amendment application in soils and generally lower yields. However, future practices should consider first maximum investment in soil conditioning and then the minimum or no-tillage management.[79] SOM is crucial to soil quality, health, the provisioning of nutrients and habitats to living organisms, soil aggregation, water storage, and create the base conditions for plant growth.[80] Human activities can change importantly natural SOM quantity and quality in soil with agricultural practices, but also by applying (natural or mineral) fertilizers.[81] This impact can be observed in all the chain (from microorganisms to herbivores).[82]

The influence of land use and management on SOM and microbial diversity is well recognized.[83,84] The SOM depletion by cultivation depending on crops nutrients demand and agricultural practices.[85] Nevertheless, it is unclear how grazing affects SOM stocks and microbial communities. For instance, Pulido-Fernández et al.[86] did not find any relation between grazing and SOM content, except under intense grazing conditions. In fact, gather a large number of animals in a reduced surface area for a short time (overstocking) in order to promote soil fertility can be an alternative practice.[43] Abdalla et al.[87] in a global meta-analysis (83 study areas) along a climatic gradient concluded that extensive grazing (within limits of carrying capacity of each area) generally decreases SOM storage, even its impact depends largely on local climatic conditions. According to Griffiths et al.[88] soils have the capacity to recover after grazing disturbance, despite the fact that their resilience is reduced where the biodiversity is low. The rooting system of many pasture species, or trees whose fruits are eatable by animals (e.g. *Quercus ilex rotundifolia* - holm oak), is extremely complex and large[89] and can store a high amount of SOM below 10 cm depth.[90] Simón et al.[91] observed also that the impact of a tree on soil fertility could reach the distance of 8 m from the trunk.[92,93]

Livestock have a strong impact on pastures floristic composition.[94] European colonizers modified American prairies from a predominance of perennial species to a higher presence of annual ones as consequence of livestock.[95] Until the first half of the 20th grazing lands (rangelands and grasslands) were rotated with other cultures such as fodder crops, at least in the Mediterranean areas.[96] The effects of these mixed practices on SOM reserves and biodiversity are still controversial. Fuhlendorf and Engle[97] for instance, propose increase landscape heterogeneity by using grazing-fire patches disturbance in order to enhance biodiversity in traditional rangelands. Contrariwise, Paustian et al.[98] consider cultivation as one of the main causes of SOM depletion in most agroecosystems.

3. Structural instability of agricultural and grazing soils

Soil structure is crucial for water retention, soil aeration and thermal conductivity.[99] It is an important indicator of soil quality and fertility.[100] Soil formation process has a strong impact on aggregation. From the point of view of agriculture and grazing this can be an advantage or not. Soils with high aggregation are more productive but at the same time are more susceptible to compaction by the machinery traffic,[101,102] rainfall impact,[103,104] addition of insecticides[30,105] and mineral fertilizers.[106,107] Soil structure is one of the most important factors for soil fertility[108] and has a strong influence on total porosity, water and air conditions, and biological properties.

Soils with high aggregation and air-filled are loose, convenient for tillage and facilitate root development.[59] Soils with stable aggregates have a good water regime, they are permeable and with a high capacity for water retention that is important for crop production. Soil stability is an important factor for thermal stability, humus mineralization, formation and fertility.[109–111]

Agricultural practices have a negative effect on soil structure,[101] and this is confirmed by several works around the globe in different climatic conditions.[112–116] Intensification and the use of pesticides and herbicides are some of the causes of soil structural degradation,[117,118] since aggregation is dependent on particle size distribution, SOM, vegetation cover, microorganisms, exchangeable cations and sesquioxides and intensive affect negatively these variables.[103] However, SOM represents the most important binding agent for forming stable aggregates.[119] The depletion of SOM caused by agriculture practices decrease importantly soil structure.[59,120] Therefore, tillage management is crucial to reduce the impacts of agriculture on soil structural stability (Fig. 3), while plowing and other tillage intensive operations have detrimental impacts.[15] The increasing area occupied by intensive monocultures, "cash-crops" (e.g. maize, sugar beet and cotton) reduces the litter input and soil structural stability.[121,122]

Agricultural practices to improve soil aggregation should consider: [1] reduced tillage; [2] mulching and residue management; [3] addition of fresh organic material (manure, compost, mulch); [4] organic fertilizers and nutrients; [5] sustainable crop management (including rotations and cover crops). These practices would reduce SOM losses and increase aggregate stability. Adopting this type of requires a deep knowledge about soils, plants and environment. Plowing disrupts soil aggregates and compact soil particles, inhibiting plant root and shoot development.[123,124] In comparison to plowed

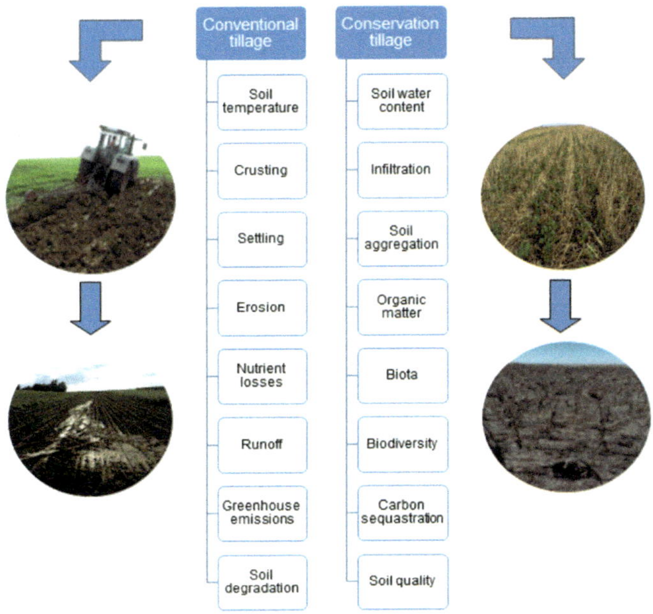

Fig. 3 Conservation and conventional management practices and soil quality indicators.

areas, no-tilled soils have more stable aggregates and high SOM.[125,126] Conservation tillage involves also reduced tillage practices and in some circumstances can be more beneficial than no-tillage, since no-tillage needs a longer period to improve topsoil structure and organic matter.[127] Reduced tillage has beneficial impacts on soil properties and reduces production costs.[103,128,129]

Beyond conservation tillage, the additional input of organic composts improves soil structure independently of the material applied. Several authors reported that organic matter increased aggregate stability, resistance to settling, and disaggregation induced by traffic and rainfall impact.[130,131] Organic matter addition in soils helps to build better soil structure by binding the soil particles with polysaccharides and bacterial exudates.[130,131] Farmyard manure increase organic carbon content, water retention, infiltration, macroaggregation, resistance to slaking and aggregate stability,[132,133] decrease soil loss, evaporation,[118,134] regulate temperature and improves soil fauna activity.[103] However, benefits from mulch depend on to the amount and quality of the residue. Composts improve soil structure and stability of the aggregates as observed in previous works,[135,136] although the impact depends also on the environmental conditions. According to de Leon-Gonzalez et al.[137] drought can limit the positive impact of compost in soil aggregation. Despite this, the effects of compost on soil structure are generally positive.

The use of fertilizers and their effects on soil aggregation have contrasting results,[138,139] mostly as consequence of the positive of these products on plant production fresh litter. Nitrogen based fertilizers increase SOM and litter production (roots and crop residues). The biomass incorporation into soil profile increases soil aggregation. Nevertheless, after several years of fertilizer application, pH decreases, affecting negatively soil aggregates,[138] although this depends also on the type of fertilizers applied. Phosphatic fertilizers and phosphoric acid can favor aggregation by the formation of aluminum or calcium phosphate-binding agents, while fertilizers rich in ammonium may act as a dispersant, like sodium.[133]

Crop management have implications on soil structure. Pagliai et al.[121] and Stagnari et al.[140] reported that crop production interventions in agroecosystems like fertilization, transition from croplands to pasture, use of cover crops, legumes and irrigation may affect positively soil structure. A meta-analysis carried out by Guo and Gifford[141] based on 74 available studies observed an increase of soil organic carbon by 19% on land converted from croplands to pasture. The same was observed when land use was converted from native forest to pasture (+8%), cropland to plantation (+18%), cropland to secondary forest (+53%). However, the inverse trend was observed in the transition from pasture to plantation (−10%), native forest to plantation (−13%), native forest to cropland (−42%), and pasture to cropland (−59%). Monocultures have the most detrimental impact on soil structure, while crop rotation including a legume (e.g. alfalfa or interseeded red clover) improves aggregate stability.[142] The increase in soil organic carbon and aggregation are directly related to the quantity of residue produced and left on the soil surface.[143] Therefore, the period between harvest and sowing in new crops, the soil should be covered with cover crops to improve SOM and aggregate stability.[144] Nevertheless, in some cases, crop rotations do not have implications on aggregate stability.[145]

According to Greenwood and McKenzie[146] livestock exerts pressure on soil comparable to agricultural machinery (e.g. a single sheep can exert a pressure between 48 and 83 kPa and a wheel tractor a pressure between 74 and 81 kPa). Proffitt et al.[147] found that damages on topsoil structure by sheep trampling at the end of the grazing period occurred regardless the grazing intensity. These damages were caused because grazing was carried out in a period with high soil moisture,[148] when the load support capacity of soil is considerably reduced.[149] Although grazing in a wet soil can imply the destruction of soil structure and compaction,[150,151] aggregates can become 2 to 4 times more stable in areas with an annual rainfall range between 900 and 1000 mm comparing to semi-arid areas (300–400 mm).[152] The most important contribution of grazing is the provision of organic manure via excreta increasing soil water holding capacity, porosity, infiltration and water stable aggregates.[138]

Long-term works carried out in Australia since 1925 observed that soils covered by permanent pastures have the smallest decline of SOM, compared to other types of agricultural practices.[153] Since soil aggregate stability depends on organic matter as binding agent along roots and fungal hyphae, the best way to conserve and improve of soil structure in grazing areas is: [1] avoiding grazing on saturated soils, [2] grazing in perennial pastures species instead of annual or cultivated ones, and [3] manage better the rotational system of the animals trying to concentrate a large number of animals in a reduce surface area (accumulation and homogenization of excreta and urine) for a short period of time.

4. Soil erosion

Erosion is a natural geological process that increased when human become sedentary (Fig. 4), approximately 7,000 years (B)C. After the establishment of agriculture, soil erosion rates became higher than soil formation.[154–156] It is estimated that soil losses caused by agricultural are 10–40 times higher than the rate of soil formation.[157] Erosion occurs when soil particles are detached as consequence of an external force such as water (in sloped areas) and wind.[158] Erosion affects especially the topsoil, the most valuable layer for crop production and vegetation development[65,159] and destroyed irreversibly 430 million ha, 30% total of the agricultural land. Natural soil erosion is responsible for lost of approximately 9 billion tons per year.[160] The effects of human activities increased exponentially soil erosion by 2.5 times (26 billion tons per year).[65,158]

Long-term studies highlighted that the most effective management to decrease water erosion is use of no-tillage or (Fig. 5) or conventional tillage perpendicular to the slope.[161] This management protects water resources from agrochemicals pollution.[162] Wendt and Burwell[163] reported that average annual runoff was about 20% less in reduced tillage compared to conventional or no-till methods. Borrelli et al.[164] observed that in Italian agricultural less favorable lands, the combined of contour farming, water collection furrows, winter cover crop and reduced tillage decreased soil loss rates in 50%. The addition of organic mulch is an effective way for reduce erosion. The application of straw mulch with 59% cover increased time to ponding and runoff generation. Median water loss decreased from 52.59% to 39.27% of the total rainfall and the median soil erosion rate

Fig. 4 Erosion in agricultural areas. Author: Ivica Kisic; Uljanik (Daruvar, Croatia), May 2010.

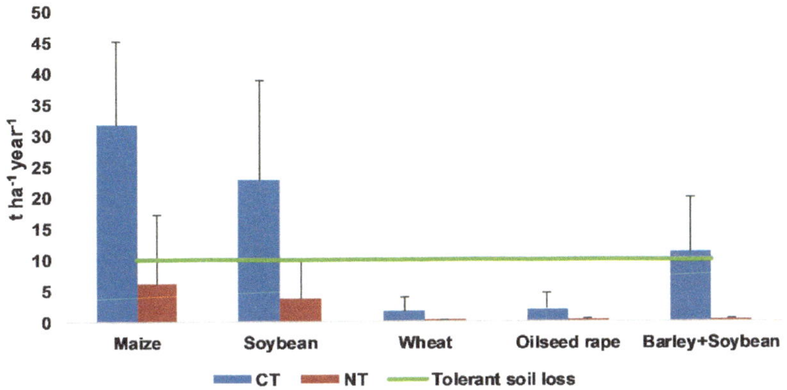

Fig. 5 Conventional and no-tillage impacts according the different crops on soil loss during 20-year research period. Each column represents the mean of same culture which appears in four occasions in 5-plot crop rotation. *CT*, conventional tillage plowing up and down the slope; *NT*, no-tillage.

decreased from 2.81 to 0.63 Mg ha^{-1} h^{-1}.[165] Zuzel and Pikul[166] reported that soil losses are reduced with the increasing straw cover ($r = -0.99$). Derpsch et al.[167] concluded that percentage of soil cover was the main factor that controlled infiltration rate. Without soil cover, the infiltration rate under no-tillage was the same as under chisel plow and conventional tillage. No-tillage in combination with adapted cover crops and crop rotations represents an efficient cropping system for soil conservation.[167]

Wind erosion occurs when there is a high wind speed, clay textured soils, poor soil structured (non-aggregated particles of clay, silt and sand) and insufficient soil protection.[168] Wind erosion is a natural process in arid and semiarid environments due to the reduced vegetation cover. Chepil[169] observed that annual soil losses by wind erosion can reach 40 t ha^{-1} and remove all topsoil. Soil tillage (Fig. 6) reduces soil cover and increases the vulnerability to wind erosion. However, this depends also on the structural aggregates size,[170] organic matter content,[171] topography[172] and meterological factors (e.g. wind speed).[65] The most evident consequence of wind erosion is the loss of topsoil.[173,174] Soil erosion costs by wind erosion are estimated to be $37.6 billion per year in the US.[175] The indirect effects of wind are the formation of dust.[176] Approximately 50% ± 20% of the atmospheric dust is attributed to wind erosion.[177] Dust emission is the greatest source of aerosols and have implications on radiation balance, climate and human and biocenoses health[176] such as respiratory problems and plants (closing of stomata).[178]

Fig. 6 Tillage induced wind erosion. Author: (A) Stjepan Husnjak, Spisic Bukovica (Croatia); February 2008; (B) Ivica Kisic, Bosansko Grahovo (Bosnia and Herzegovina); September 2012.

Soil loss due to root crop harvesting (SLCH) is one of the least known forms of erosion, despite the fact that is recognized as an important degradation process.[179] This occurs as consequence of the extraction of sugar beet (Fig. 7), potato tubers and some other root crops such as biscuits, onions, carrots, roots, asparagus, batat, celery and peanuts.[180] Previous works observed that the highest soil loss was recorded during sugar beet harvest,[181] followed by potato,[182] cassava,[183] yam or sweet potato.[184] In all cases, soil losses did not exceed 20 t ha^{-1}.[185,186] The dominant factor that affects soil losses due to crop harvesting is the water content at the time of harvest (Fig. 8),[187] although texture is also an important factor to consider. Swelling processes in soils with high clay content (soil fraction that holds large amounts of water) increase cohesion and a stronger adhesion to the roots.[186]

Soil water (sheet) erosion and compaction are the most important degradation process in grazed lands.[51] Soils losses are particularly high at the

Fig. 7 Root shape influences soil loss after harvesting. Author: Ivica Kisic; Ovcara (Vukovar, Croatia); October 2008.

Fig. 8 (A) Climatic conditions during sugar beet: (A) harvested sugar beet on water-logged soil; (B) high soil loss on harvested sugar beet in wet conditions. Author: Juergen Eggers; Zdenci (Orahovica, Croatia); October 2010.

beginning of the rainy season (autumn) after the grazing period and the existence of large areas without vegetation cover.[15,188] In concave areas, the development of gullies is frequent.[189] Rill erosion is also an important process, especially in rotated cropped areas.[190] Fig. 9 shows some evidences of water erosion in Mediterranean rangelands such as lateral gully formation (Fig. 9A) due to continued pass of animals. Root exposures in a holm oak

Fig. 9 Examples of water erosion in Mediterranean rangelands: (A) Lateral gully, (B) Root exposure by sheet erosion, (C) Rill erosion and (D) Gully erosion.

tree caused by sheet erosion processes (Fig. 9B). Fig. 9C shows rills formed in a rotational fodder crop (*Avena sativa*), Figure D a gully of 3 m width formed as consequence of Iberian pigs overgrazing.

Sheet erosion is an important process in grazed lands, nevertheless gully erosion produces more sediment. Schnabel et al.[191] reported average losses of 0.63 t ha^{-1} yr^{-1} (range: 0.12–1.34 t ha^{-1} yr^{-1}) by sheet erosion in Mediterranean rangelands, while the losses by gully erosion were high (1.55 t ha^{-1} yr^{-1}, with a highest peak of 5.57 t ha^{-1} yr^{-1}). However, the highest rates of soil erosion are observed in fodder crops of *A. sativa* (rates above 100 t ha^{-1} yr^{-1}), where rills were formed as a consequence of a single extreme rainfall event.[190] These erosion rates are low when compared to the observed in agricultural areas,[192] where erosion rates largely exceed exceed pedogenesis degree. These values are always higher than the observed under forest cover. Sala[193] observed an average soil loss of 0.05 t ha^{-1} yr^{-1} in a Mediterranean forest dominated by holm oaks (tree used for feeding livestock in the dehesa/montado system) in NE Spain.

The main driving factor on water erosion in grazing (and agricultural) areas is the lack of soil protection. Schnabel et al.[191] observed that soils with large vegetation cover (>60%) have minimal erosion (0.30 t ha^{-1}) even after extreme rainfall events while with a vegetation cover below 20% soil losses increase significantly. In this context, Pulido-Fernández et al.[86] found a logarithmic relationship between animal stocking rates and the percentage of surface area occupied by bare patches (Fig. 10). Grazing intensity is crucial to understand soil erosion processes in rangelands. Pulido

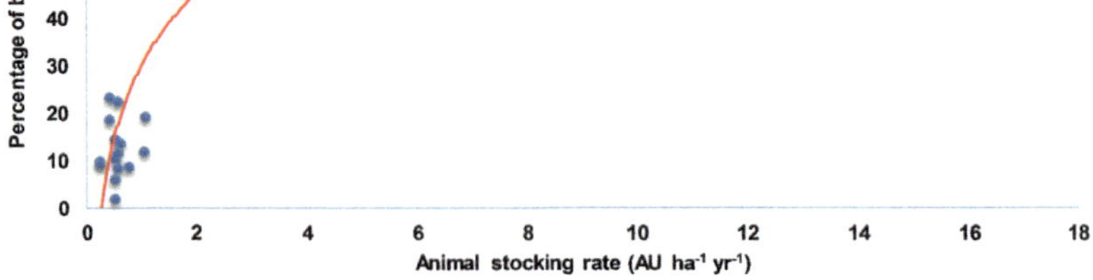

Fig. 10 Scatterplot of the percentage of bare soil against animal stocking rate in a wet year.

Fig. 11 Pulverization and wind erosion provoked by Iberian pigs in a Spanish rangeland. Author: Manuel Pulido Fernández; Aliseda, Cáceres, SW Spain; September 2017.

et al.[51] suggested a maximum level of animal stocking of 1.0 AU ha^{-1} and 0.5 AU ha^{-1} in wet and dry years, respectively.

There is a lack of studies focused on the effects of grazing in wind erosion. In grazing lands has been so far a very little studied phenomenon. Animals can increase soil pulverization under dry and windy conditions (Fig. 11).

Wind erosion in arid environments is an important phenomenon, especially in fine textured soils. Zhao et al.[194] considered that wind erosion induced by heavy grazing as one of the most important causes of desertification after they observe a loss of 25 cm of soil in 5 years in Inner Mongolia. Hoffmann et al.[195] identified also in Inner Mongolia a loss of 4.8 and 0.4 kg ha^{-1} of SOM and nitrogen due to wind erosion and heavy grazing. It means heavy grazing can provoke soil loss and particularly of fine particles linked to its fertility.

5. Soil compaction

Soil compaction is one of the most important evidences of soil degradation,[101,196] and undermines soil quality and health, reducing their potential for agricultural production. The ideal soil condition contains 50% of solid particles (90% mineral and 10% organic), and 50% is porous, where half is filled with water and the other half with gas. This ratio of solid and gas is conditioned by soil structure. Any kind of management activities involving machinery reduces soil aggregation, porosity, and increases bulk density and soil compaction (Fig. 12A). In case of conventional tillage

Fig. 12 (A) Waterlogging due to traffic-induce compaction (left); (B) poor plant population on compacted land. Author: Igor Bogunovic; Potok, Croatia, 2013.

involving multiple tillage passages, the plowed layer is loosened and recompacted every year.[59] Compaction increases soil strength,[197] destroy aggregates,[101] reduces the infiltration[198] and fertility, leading to additional fertiliser requirements and increasing production cost.[148] After the wheel traffic topsoil pores are mostly destroyed,[199] reducing also gas exchange between soil and atmosphere.[200] This has implication on soil chemical properties.[201] Increased compaction inhibits soil microbiological activity,[202] reduces the number of earthworms[203] and other living organisms in the soil, decreasing crop yields.[204] Root development is also reduced,[205] and increases the emission of greenhouse gases.[101] Soil compaction increases the probability of weed infestation.[204]

Soil compaction affects negatively all soil properties. Reduces crop yields (Fig. 12B), increases fuel consumption, and machine breaking. Soil aeration is poor and aerobic microbiological processes are slow reducing the rate of organic matter transformation.[206] Soil compaction also affects the mineralization of organic carbon and nitrogen,[207] as well as the concentration of CO_2 in the soil.[208] Since soil fauna is strongly affected by compaction, the rates of mineralization are reduced. Soil compaction is harmful for soil in rhizosphere, especially in semiarid and arid areas. Compaction can be beneficial in light textured soils, since enhances contact of soil and seed increasing the germination process. In sandy soils, compaction reduces the size and continuity of the pores that are too large to retain water. Therefore, in these types of soils, compaction may have a positive impact in plant germination.[209,210]

In 1971, reduced crop yields in the USA as consequence of compaction were estimated in 1% in the north to 10% in the south. Worldwide it is estimated that 68 million ha of soils are compacted as result of mechanization.[211]

This type of land degradation occurs on 4 million ha in the "wheat belt" in Western Australia and 33 million ha in Europe.[212,213] In Europe, compaction is responsible for 17% of the total degraded area.[214] Moreover, extent of compacted soils in agricultural areas of Central Europe are Bulgaria (47%), Croatia (25%–35%), Czech Republic (40%–50%), Hungary (30%–35%), Poland (20%–25%) and Romania (54%).[59]

Soil compaction is a result of machine traffic, weight and used tools. These impacts can be act alone or coupled and are especially adverse when the soil is moist. Tillage effects on compaction are attributed to poor management, where farmers cultivate year by year at the same depth creating an impermeable tillage pan. Wheel pressure on the open furrow contributes to compaction of soil on the edge between tilled and undisturbed soil. In this type of management soils have several compacted layers within the profile. According to Birkás et al.[205] several years of plowing with a mouldboard or disc plow creates poorly permeable layers on the border of the tilled layer, which increase in the depth and time. Tillage-pan has multiple negative effects. In wet period this layer become quickly saturated as consequence of water availability and infiltration, while in dry period tillage pan prevent water to move to rhizosphere.

Traffic-induced compaction depends on the intensity, pressure per area, number of passes, duration of the impact, size of the contact area between soil and tire and the soil characteristics.[101,215] Efforts to reduce the wheel traffic on the soil, however, the use of heavy machinery makes all these efforts useless. The number of large tractors weighing more than 10 tons is increasing. The heaviest harvesters can weight more than 25 tonnes and the liquid organic fertilizer spreaders can weigh over 30 tons. Sugar beet harvester weighs about 35–40 tons (two-axle), while three-axle model can weight more than 50 tonnes when is fully loaded.[216,217] When the wheel presses the soil, the extent of compaction in the topsoil is determined by the wheel pressure and the area of implement. Subsoil compaction mostly depended by the wheel load, number of traffic events and soil water content.[218] Since the subsoil usually has more strength than topsoil, its physical conditions are less affected by traffic. Nevertheless, repeated traffic can cause cumulative compaction. This negative impact can be greater when machinery is heavy, although soil compaction was also identified using light machinery.[197,219] Birkas et al.[59] observed in wet soil conditions that a four-tonne axle load created a compacted layer at 30 cm in depth. The depth of the compacted layer can increase until 50 cm if traffic is done with a ten-tonne axle load. The most important impact is produced in the first passage,[101] increasing with number of passes,[220] irrespectively to machinery weight.[221,222]

Soils differ in their sensitivity to compaction.[148] Medium textured soils are more vulnerable to compaction than sandy soils.[223] Soils rich in silt are more susceptible to compaction than sandy or clayed soils.[101] Brais and Camiré[224] and Ampoorter et al.[225] found that sandy soils can be also prone to compaction. In coarse textured soils, wheel loads generated a vertical preferential direction of pressure, while in soils with finer soil texture the propagation of the pressure is multidirectional.[226]

There are several ways to reduce traffic negative consequences on soil. First, soil tillage system or machinery design can be changed, while crops with greater tolerance to compaction could be predominant in crop rotation. However, the possible solution is not developed, since today traffic impact depends on climate, soils and cropping systems. Usually, recommendations involve agro-technical operations like: subsoiling, reducing tillage, controlled traffic farming and increase of organic matter content in soils. There are several simple measures that can reduce the compaction such as adjusting the type inflation pressure to the loading and driving situation and choosing suitable days for field operations when the soil is dry.[219] Other measures include using no-tillage management in crop production, using efficient wide-spread machines and tools, limitation of the axle load of machines, use of dual wheels, tracked vehicles, permanent traffic lanes for field transport, coupling several agro-technical operations in one pass and drainage of soil, avoiding plowing and disking operations at the same depth.[59,101,148] Loosening the soil by subsoiling or ripping partially mitigates the negative effects of traffic. The residual effect of loosening is short and last only one or two seasons.[227] Soil compaction is one of the central problems in intensive agricultural production. Compaction reduces yield and soil nutrients.[59] In addition to the construction of tools and machines, it is necessary to have a correct soil management that reduces the negative effects of traffic. It is also crucial to develop soil tillage and cropping system adapted to local environment.

Soil compaction is an important form of land degradation process in grazed areas. Compaction has been attributed to cattle grazing in areas with drainage problems (e.g. poaching), although paradoxically most of the work was carried out in arid and semi arid areas of Spain. Blanco-Sepúlveda[228] found that goat grazing can induce soil compaction in Mediterranean mountainous. This is a consequence of the pressure exerted by animals and their hooves on the ground.[146] The problem is when this occurs in soil rich clay during the winter period, when the soil is moist. Greenwood and McKenzie[146] estimated that the pressure exerted by livestock on ground depends on animal weight, hooves size and kinetic energy. Australian sheep induces a pressure on the soil of 48–83 KPa, adult cattle, 98–192 KPa, goats, 60–73 KPa, horses, 54–95 KPa and camels 27–40 KPa, respectively. Kangaroos (42–92 KPa) had values of pressure similar to a person (41–108 KPa, depending on shoes). Bilotta et al.[150] found that animals in intensive agriculture farms located in poorly drained soils caused four different types of compaction: trampling, treading, pugging and poaching.

The effects of soil compaction at short, medium and long-terms in grazing areas are similar to the identified in agriculture areas. Pulido et al.[51] found an increase of soil bulk density from 5 to 10 cm in depth in Mediterranean rangelands heavily grazed (animal stocking rate: 1.0 cow ha^{-1} yr^{-1}) and consequently a significant reduction in pasture production and nutritional value (noticeable reduction of rich in protein legumes favoring the growth of other less nutritive and palatable grasses). Vegetation has difficult to recover in frequent grazed areas.[90] Greenwood and McKenzie[146] found that trampling could affect the soils until 15 cm depth, while Pulido et al.[51] found affectations until 10 cm. Pulido et al.[90] reported values of penetration resistance above 2 MPa in the 70% of the measurements carried out below 5 cm, especially in areas affected by heavy grazing (Fig. 13).

Soil compaction is mainly measured with bulk density and soil penetration, while soil penetration resistance is usually used to assess plant rooting

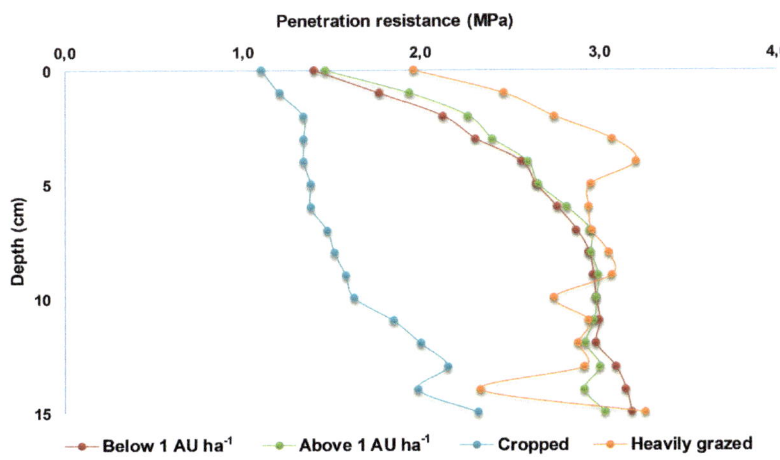

Fig. 13 Variability of soil penetration resistance in depth comparing different land management.

Fig. 14 Occurrence of animal paths (permanent bare soil) caused by soil compaction in a Spanish rangeland (Deleitosa, Cáceres, Spain). Author: Javier Lozano; Cáceres, Spain; February 2008.

capacity.[148] Soils with problems of compaction become unproductive, and one of the evidences is the large areas affected by bare soil that recover with difficulty to trampling (Fig. 14). The areas most affected by trampling in a farm are the ones located near pond, corrals or entrances to fenced areas.

6. Soil crusting

The occurrence of soil crusts, i.e. the formation of a hard surface on a bare soil surface, is a natural process enhanced in many cases by anthropic effects due to land mismanagement. Ultraviolet rays' incidence in bare soils decrease microbiological activity. Sun direct radiation also results in water losses and high evaporation,[229,230] deteriorating soil physical properties.[231] The kinetic energy of raindrops leads to disaggregation of soil aggregates and dispersions of soil colloids.[232] After drying, crusts are formed. Surface crust reduces water infiltration and increases runoff and soil losses.[233] Soil surface crust has normally 1–3 cm thickness.[59] The presence of crusts also reduces crops germination.[218]

Soil crusts are formed especially in soils poor in organic matter. Nevertheless, several types of crusts are recognized[234]: (1) physical and biological crusts - dominant in semi-arid and arid regions; (2) biological or micro-biotic crusts, formed by algal growth; (3) chemical crust created due to salt accumulation on soil surface; (4) structural crust - formed due to the raindrop impact, irrigation and physiochemical dispersion of clays.[234,235]

Šimansky et al.[231] observed a negative effect of clay content on crusting and a positive impact with the increase of water content.

Surface crusts are created by traffic on the wet ground or by using inadequate tools during management.[236] Although crusting can be easily destroyed by frost, wetting and drying cycles and by tillage interventions, crusts can cause considerable reduce plant germination, water infiltration and gaseous exchange.[237] In such conditions carbon dioxide accumulates in the soil, and oxygen concentration is limited for aerobic respiration and seed germination.[238] To destroy soil crusts is necessary wet the soil or removed mechanically, especially in the vegetation period (Fig. 15). Soils can be shallowly tilled with a harrower and a rotaspa spading machine.[59] The presence of stable aggregates reduces soil crust.[239] Therefore, the addition of gypsum, calcium carbonate and organic amendments like farm-yard manure; crop residue (leaves, straw) and husk and soil conditioners e.g., certain organic compounds, polyvalent salts and various synthetic polymers contribute to reduce crusting.

Soil (bio) crusts are also detected in grazing lands (Fig. 16). They usually occur where soil is bare and exposed to weathering processes for a long time as consequence of fire, drought, crops and overgrazing.[84] After this disturbance, soils are firstly colonized by cryptograms (nonvascular plants) species such as lichens or mosses. Also, physical crusts can occur in semi-arid environments.[240] In some circumstances, soil crusts are beneficial in desert[241] but in humid areas are a consequence of poor management.[242] In (hyper) arid environments livestock disturbance is negative to soil crusts (e.g.[243,244]). Normally, livestook trampling increases soil erosion and decrease water infiltration. Nevertheless, some studies found different results. Eldridge and

Fig. 15 (A) Poor aggregates soils are susceptible to surface crust and (B) should be mechanically cultivated in order to provide gaseous exchange between soil and atmosphere. Author: Ivica Kisic; Krivaja (Orahovica, Croatia); June 2011.

Fig. 16 Soil crust formed in areas used by animals for resting in summertime days (tree shadow) in a Spanish rangeland (Monroy, Cáceres, SW Spain). Author: Manuel Pulido; Cáceres, Spain; September 2010.

Greene[245] found in Australia that, biocrusts increased water infiltration capacity of soils and had a high capacity for nitrogen fixation. Graetz and Tongway[246] observed that the infiltration capacity in soils under heavy grazing, where biocrusts had been broken, was three-fold than in ungrazed areas. Muscha and Hild,[247] found that the presence of biocrusts as an indicator of healthy landscapes, although lichens are sensitive to grazing effects.[248] In sub-humid and wet environments, there are a lack of studies since they are not a widespread phenomenon. For instance, local ranchers of dehesa/montado system are only worried about the occurrence of crusts in their soils because they are areas where pasture cannot growth (personal communication: Casto Herrero and Pedro Flores, dehesa landowners).

7. Acidification

Acidification represents one of the most important chemical soil degradation processes. According to Oldeman,[249] these negative degradation processes by excessive acidity affects occurs on 6 million of hectares. Soil acidity in sub-humid and humid environments represents a serious problem for plant development due to toxicity caused by the high concentrations of aluminum and manganese in the rhizosphere.[250] Soil reaction (acid, neutral or alkaline) is determined by the concentration of hydrogen and hydroxide ions in soil solution (e.g. at pH 7). Soil pH is the negative logarithm of the hydrogen ion concentration. The reaction of the soil solution in different soils varies from pH 2 to 10.5, the activity of hydrogen ions is between 10^{-2} and $3\ 10^{-11}$ mol L^{-1}.

Soil acidity has an important influence on the nutrient dynamics and SOM mineralization. According to Foy,[251] soil pH determines the type, number and activity of the microorganisms. By correcting the pH to a neutral value, the absorption of nitrogen, phosphorus, sulfur and microelements by plant roots is facilitated. If pH is below 5.5, fungi are the most active organic matter decomposers, while at pH over 6.0 actinomycetes and bacteria are the main decomposers, although these microorganisms have different propagation optimums with respect to soil acidity.[252] pH is not so critical for soil ammonification process compared nitrification, which is importantly reduced if the pH is below 6.0 or above 8.0. At a pH below 4.5, nitrification is strongly reduced.[253,254] One-half to two-thirds of the total phosphorus is in the organic form and most of the organic-bound phosphorus is created from dead microorganisms. The availability of sulfur depends on the intensity and speed of SOM mineralization.[255] Microbiological activity affects importantly the solubility and oxidation of manganese, zinc, copper, aluminum and molybdenum.[256]

Soils become acidic when the majority of the exchangeable cations are composed by hydrogen and other forms of hydrated aluminum. Although some acidic soils develop from the acidic parent material, most are formed as consequence of leaching. The infiltration of water rich in hydrogen ions from various weak acids (e.g. carbonic and organic acids) replaces the cations such as calcium, magnesium, potassium and sodium.[109]

Summarizing, the causes of soil acidity: [1] soil was developed in acidic parent material; [2] major cations are washed under the influence of rainfall; [3] roots excretes hydrogen ions as acids; [4] the presence of nitrogen, since most of their forms is acid; [5] the application of sulfur-containing fungicides creates acidic conditions in the soil and [6] decomposition of organic matter releases hydrogen.[218]

Soil acidification is a natural process, but it can occur as well as consequence of agricultural and grazing activities. The depletion of calcium, magnesium and potassium by crop harvest and the application mineral and organic fertilizers of increases soil. Higher the application rate, higher the rate of acidification.[257] Mineral fertilization increases the input of nitric and sulfuric acid, although phosphorus fertilisers like monocalcium phosphate releases hydrogen ions.[258] Previous works identified a negative impact of the use of mineral fertilisers on soil pH. Barak et al.[259] observed that after 30-years of fertilization, the soil became more acid, while cation exchange capacity, base saturation, and calcium and magnesium ions decreased with the application of ammonium-based fertilizer. Similar results were reported elsewhere.[260,261] Ammonium-based fertilizers have a negative impact on soil quality, since they increase the leaching of base cations and create nitric acid.[262] In addition, inputs of sulfur increase soil acidity after oxidation. Normally, farmers are encouraged to add sulfur for plant nutrition. With the exception of poultry manure, other types of farmyard manure are better for the soil, since is composed by organic residues rich in calcium.[218]

In industrial cropland farms, farmers use green manure as a source of organic matter. However, after continuous application of green manure, the decomposition of SOM increases the emission of CO_2 and increase soil acidification,[218] regardless the content of nitrogen-fixing plants. However, nitrogen-fixing species also contribute to soil acidity as they fix atmospheric nitrogen and form ammonium.[263]

Proper soil management for acidic soils include addition of calcium rich materials. Liming is a positive method to alleviate soil acidity. Lime is mainly composed by carbonates, bicarbonates, oxides and hydroxides of calcium or magnesium.[264] However, they effect of lime depends on soil properties, mainly chemical composition and texture.[265] Lime applications influence aggregate stability and other soil physical properties like bulk density and infiltration.[138] Liming increase crop yields,[266] producing a high amount of biomass that return to soil, increasing SOM and soil aggregates.[267] The application of lime changes the chemical composition of soil solution. Aluminum and manganese toxicity is suppressed, while calcium and magnesium increase. In addition, the increase of pH, ionic strength and ionic composition reduce clay dispersion particles.[138] The addition of lime depends on soil clay content, cation exchange capacity and base saturation, since these determine soil buffer capacity.[264] This is reason why two different soils with similar pH need different lime application rates. There are a several methods for lime requirement determination methods that can be calculated online (e.g. https://aglime.org.uk/lime_calculator.php).

Livestock modifies chemical fluxes and pH in soil by ingesting nutrients and returning them via excreta. The effects of grazing on soil pH are controversial and depend very much on the studied site. For instance, Martins et al.[268] found a reduction of acidification in mixed-systems of southern Brazil (humid subtropical temperate climate type) as consequence of grazing. On the contrary, Dorrough et al.[269] addressed the necessity of reduce grazing to decrease soil acidification in temperate grasslands of Australia. Peters et al.[270] highlighted the importance of the management that can affect soil nitrogen contents, since can induce soil acidification, especially as consequence of SOM and animal's urine that releases nitric acid. In addition, the removal of vegetation is accompanied by a washing of major cations[188] and the fixation of nitrogen by legume pastures (because they are considered pastures of higher nutritional value) can alter the natural chemical balance of soil.[271]

Several solutions have been proposed to reduce soil acidification in grazing environments, despite the fact that they are not serious as in agricultural areas. Bell et al.[272] suggested the progressive replacement of annual species of pastures by perennial with a more stable root system. Martins et al.[273] encourage the use of animals in mixed-systems even after having applied lime amendments because they reduce soil acidity in the deeper layers.

The application of lime to reduce acidification is a well know in agricultural areas. Nevertheless, is not a realistic solution in rangelands or grasslands larger than 100 ha. In properties of this dimension, animal selling is not abundant. In addition, many of these grazing lands are explored areas where soil is shallow and reduced fertility (for this reason they are not destined to commercial crops) developed in siliceous rocks (e.g. schists).[274] Therefore, a good way to guarantee the sustainability of these systems is to reduce soil erosion and compaction and to manage pasture species well adapted to acidic conditions. It is important to favor perennial species with strong root system that contributes to soil fertility at a medium or long-term (personal interview with a rancher member of the Savory Institute).

8. Salinization and alkalization

Soil salinization and alkalization are processes recognized to be a severe environmental hazard. Several works stated that soil salinity is an important abiotic stress that severely impact agricultural production, sustainability,[275–277] and can affect human health.[278] Soils are naturally salt in arid and semi-arid regions where precipitation is lower than evaporation. Two types of salinization are recognized: primary (salt accumulation through natural processes) and secondary (human-induced and ever-rizing groundwater). The presence of salts in the soil can derive from parent material, irrigation, fertilizers, manures, composts or other soil amendments. Sodicity, as a secondary result of salinity, is a consequence of the high accumulation of sodium and have an important impact on clay dispersion.[279] According to Metternicht and Zink[280] approximately 95 million ha of soils are naturally salty, while 77 million ha are salty as consequence of human management. This represents 20% of the world's irrigated lands.[281] The numbers about the areas affected by high salt content are not consistent and differ among studies.

Salts are originary from bedrock weathering while the groundwater is high (up the 1 m), like is noted in India,[282] Turkey[283] or Croatia.[284] Evapotranspiration reduces soil water, and the salts start to accumulate in soil that can result in serious consequences for plants and animals.[285] However, salt-affected soils may develop along seacoasts or river delta regions where seawater penetrates the rivers,[286] aquifers[287] or groundwater in deeper inland used for irrigations. Several works in river deltas like Yellow River,[288] Nile delta,[289] Neretva delta[290] observed an increase of saline ions: sodium, potassium, calcium, magnesium and chloride. Agriculture is recognized as source of the salinization and alkalization processes. Irrigation affects soil salinity, when the water is rich in dissolved salts. Excessive application of irrigation water can increase water table, and waterlogging, restricting crop plantation and productivity.[291] Secondary salinization could be created also by other human activities like deforestation,[292] accumulation of air-borne and water-borne salts,[293] chemical contamination[294] or overgrazing.[295]

According to Richards[296] soils are classified as salty by their electrical conductivity of the saturated extract (ECe), exchangeable sodium percentage (ESP) and sodium adsorption ratio (SAR), and soil pH (Table 1).

Salt damages depend on species, variety, growth stage, environmental factors, and nature of the salts.[279] The major problem of saline and sodic soils occurs when sodium, a very soluble cation that do not bound easily onto soil particles, start to accumulate and dominate over the calcium. It causes soil structural deterioration such as slaking, swelling and clay dispersion leading to compaction[297] and the consequent decline of soil productivity. Compaction can be observed by the creation of a surface and sub-surface crust.[113]

The reduced vegetation growth in salt affected soils[298] results in low carbon inputs and degradation of physical and chemical properties.[299] Sodic soils affect plants nutritional imbalances including a deficiency of calcium.

Table 1 Characteristics of saline and sodic soils.[296]

	Saline	Sodic	Saline-Sodic
EC	>4	<4	>4
ESP	<15	>15	>15
pH	<8.5	>8.5	<8.5

EC, Electrical Conductivity (dS m^{-1}); *ESP*, Exchangeable Sodium Percentage.

Several authors[300–302] observed that salt accumulation in soils have impacts on plant metabolism, vigor, and production resulting in lower yields. Due to their low production value, saline and sodic soils require restoration measures and maintain soil pH between 5.5 and 8.2 and increase exchangeable calcium content.[303] In several cases, improvements are carried out by loosening the soil and break salt crusts.[59] Biological amelioration can be carried out by applying green manure in crops.[304]

Reclamation strategy for saline soils consists of leaching chloride and sulfate with low electrical conductivity water in order to reduce salinity in rhizosphere and reduce the migration of salts to groundwater.[297] Successful reclamation of saline-sodic and sodic soils implies the reduction of sodium from the cation exchange complex. This could be achieved by applying dissolved limestone or gypsum. Gypsum is the most common material applied to increase calcium that replaces sodium in colloidal surface. Sodium with sulfate forms sodium sulfate, which is more soluble in water and is easily leached. After gypsum application, it is recommended to use non-saline water for leaching sodium bellow the root zone. Gypsum increases infiltration rate and decrease of pH, EC, and ESP[275] improving soil stability.[305] Hanay et al.[306] demonstrated that the application of gypsum (50 t ha^{-1}) decreased the ESP from 45% to 7% under laboratory conditions. In Egypt, Abdel-Fattah et al.[307] recommends use of fine of gypsum particles increasing the effectiveness in reducing salinity and sodicity. Ahmed et al.[308] recommended a maximum fresh (19.79 t ha^{-1}) and dry yield (4.43 t $ha^{-}1$) in fodder production, when gypsum was applied at 100% gypsum requirement. Rasouli et al.[309] found a significantly higher yield of wheat at treatment with 100% soil gypsum requirement compared to treatments with 50% and 75% of gypsum requirement application rate. The yield was higher in soils treated with coarse gypsum compared to the plots where fine gypsum was applied. On the other hand, Bennett et al.[310] found no significant differences in wheat and cotton crop performance after gypsum amendment. Luo et al.[311] recommended the application of gypsum during several years to complete amelioration of hard saline-sodic soils.

For reclamation of sodic and saline-sodic soils, especially in carbonated soils, elemental sulfur addition may increase calcium indirectly. Sulfur occurs in organic (>90%) and inorganic (SO_4^{2-}) forms and is an essential nutrient for crop and pasture growth, but can increase stress tolerance in plants.[312] To be effective in reclamation of sodic and saline-sodic soils, sulfur should be oxidized to sulfuric acid, which reacts with the calcium carbonate to form gypsum. This reaction (oxidation) of elemental sulfur is slow and depends on soil moisture conditions and microbiological (genus *Thiobacillus*) activity. Sodium is replaced by calcium on the cation exchange complex, increasing soil hydraulic conductivity increases and sodium salts leaching. Positive results in sodic reclamation were observed.[313–316] Soil pH was reduced and the conditions for nutrient uptake were improved. Similar results were identified in soils amended with sulfuric acid,[317] pyrite,[318] aluminum sulfate and calcium chloride.[319]

Another favored amendment for reclamation of sodic and saline-sodic soils is farmyard manure (FYM). The addition of FYM promotes microbiological activity.[320,321] The created carbonic acid, acidifies soils; increase the mobility of phosphorus and transforms calcium carbonate, contributing to soil colloids coagulation of colloids and structure stability.[322] Positive impact of FYM on decreasing soil pH in sodic soils is confirmed in many studies all around the World.[306,308,323,324] However, FYM and other organic amendments have a limited application as consequence of the high cost, low efficiency in plant nutrition or pollution.[324] Organic matter can accelerate sodium leaching, increase water infiltration, water-holding capacity, carbon and nitrogen contents, enzyme activities and aggregate stability and decrease the bulk density, ESP and EC.[135,294,325]

Livestock grazing does not affect soil salinization and alkalinization as agricultural practices (e.g. irrigation). Wang et al.[295] observed that overgrazing is one of the main drivers of water evaporation in wetlands. Shang et al.[326] recommended the reduction of animal stocking to 25% in order to avoid salinization and alkalinization. While the effects on soil salinization are more visible in arid areas (evapotranspiration > precipitation), this phenomenon has been also observed in temperate climate as consequence of grazing. Cisneros et al.[327] identified soil salinization in Córdoba province (Central Argentina), an area with an average annual rainfall of 725 mm yr^{-1}. Vegetation removal by grazing enhances evaporation, reducing considerably water infiltration and leaching of salts. In North Africa, excessive irrigation increasing salinization as consequence of poor management, decreasing the fertility of large extensions of land.[328] Salinization increases the area with bare soil, since the high content of salts make soil toxic to some plants. Despite this these areas are still grazed. In this context, grazing can be considered as cumulative impact in soil salinization process.

One solution to remediate salinization is planting several hectares of land using species well adapted to arid conditions and saline soils such as *Atriplex nummularia*. Nevertheless, the effects of these practices can be useful in a period between 4 and 12 years after the plantation, since *A. nummularia* does not survive longer. Therefore, it is necessary to prevent premature grazing to avoid disturbances in plant growth.[329]

9. Conclusions

Soil degradation is a global phenomenon that affects millions of hectares of agricultural areas. It is a consequence of organic organic matter depletion, soil erosion and compaction, crusting, acidification, alkalinization and salinization. Organic farming and sustainable agriculture are some of the new practices and concepts that can reduce land degradation and increase soil quality and crop production. Agricultural intensification and overgrazing are considered as the main drivers in the reduction of soil organic matter and lost of biodiversity. Some alternatives such as (semi) permanent cover, minimum tillage and no-tillage crops as well as holistic grazing management and the design of mosaic-shape landscapes have been fruitfully checked in different ecosystems (e.g., tropical, Mediterranean, etc.). This lack of organic matter as binding agent reduces considerably soil aggregation, i.e. the possibility of soil to store water and air. Soil erosion is by far the most studied degradation phenomena since it is the cause of the destruction of 30% of the total agricultural land. No-tillage is considered as the most effective management to reduce water erosion. Wind erosion, by its part, is more frequent in (semi) arid areas in which to keep a minimum a vegetation cover is challenging since water scarcity is an important constraint. The use of machinery, particularly heavy tractors, is compacting the topsoil and generating an impermeable tillage pan. In addition, the pressure exerted by a domestic animal can be comparable to a caterpillar machine. These compacted soils are frequently decompacted by using this kind of machinery but this practice is, on one hand, inducing recompaction and, on the other hand, is destroying soil crusts that can be beneficial in (semi) arid environments. The use of mineral fertilizers and the excess of manure are provoking problems of acidification that can be only reversed with lime amendments, which can be a no realistic solution for many farmers due lime high price. Finally, other problem that is progressively affecting more people is salinization, particularly in (semi) arid areas where water management for irrigation is not the adequate. It is converting a large surface of fertile lands in bare areas where only halophytic species can survive.

Acknowledgments

This work was supported by the Croatian Science Foundation through the project "Soil erosion and degradation in Croatia" (UIP-2017-05-7834) (SEDCRO).

References

1. United Nations. *World population prospects: the 2017 revision, key findings and advance tables*. Department of Economic and Social Affairs; 2017. Population Division Working Paper No. ESA/P/WP/248.
2. Fereres E, Orgaz F, Gonzalez-Dugo V. Reflections on food security under water scarcity. *J Exp Bot* 2011;**62**(12):4079–86.
3. Unnevehr LJ. *Food safety in food security and food trade. 2020 Vision for food, agriculture and environment*. Washington, USA: International Food Policy Research Institute; 2003.
4. Dudeja P, Singh A. Food safety issues related to plant foods at farms. In: Gupta RK, Dudeja P, Singh Minhas A, editors. *Food safety in the 21st Century*. London, United Kingdom: Academic Press; 2017. p. 179–91.
5. Gomiero T. Soil degradation, land scarcity and food security: reviewing a complex challenge. *Sustainability* 2016;**8**(281).
6. Stoate C, Boatman ND, Borralho RJ, Rio Carvalho C, de Snoo GR, Eden P. Ecological impacts of arable intensitification in Europe. *J Environ Manag* 2001;**63**:337–65.
7. Cowie AL, Orr BJ, Castillo Sanchez VM, Chasek P, Crossman ND, Erlewein A, Louwagie G, Maron M, Metternicht GI, Minelli S, Tengberg AE, Walter S, Welton S. Land in balance: the scientific conceptual framework for land degradation neutrality. *Environ Sci Policy* 2018;**79**:25–35.
8. Peng W, Berry EM. Global nutrition 1990–2015: a shrinking hungry, and expanding fat world. *PLoS One* 2018;**13**(3):e0194821.
9. Chen C, Xu X, Yan Y. Estimated global overweight and obesity burden in pregnant women based on panel data model. *PLoS One* 2018;**13**.
10. Godfray HC, Crute IR, Haddad L, Lawrence D, Muir JF, Nisbett N, Pretty J, Robinson S, Toulmin C, Whiteley R. The future of the global food system. *Philos Trans R Soc Lond B Biol Sci* 2010;**365**:2769–77.
11. FAO, IFAD, UNICEF, WFP, WHO. *The State of Food Security and Nutrition in the World 2018. Building climate resilience for food security and nutrition*. Rome: FAO; 2018.
12. von Braun J, Rosegrant MW, Pandya-Lorch R, Cohen MJ, Cline SA, Brown MA, Soledad Bos M. *New risks and opportunities for food security scenario analyses for 2015 and 2050*. Washington: International Food Policy Research Institute; 2005.
13. Pretty J, Sutherland WJ, Ashby J, Auburn J, Baulcombe D, Bell M, Bentley J, Bickersteth S, Brown K, Burke J, Campbell H, Chen K, Crowley E, Crute I, Dobbelaere D, Edwards-Jones G, Funes-Monzote F, Charles H, Godfray J, Griffon M, Gypmantisiri M, Haddad L, Halavatau S, Herren H, Holderness M, Izac AM, Jones M, Koohafkan P, Lal R, Lang T, McNeely J, Mueller A, Nisbett N, Noble A, Pingali P, Pinto Y, Rabbinge R, Ravindranath NH, Rola A, Roling N, Sage C, Settle W, Sha JM, Shiming L, Simons T, Smith P, Strzepeck K, Swaine H, Terry E, Tomich TP, Toulmin C, Trigo E, Twomlow S, Vis JK, Wilson J, Pilgrim S. The top 100 questions of importance to the future of global agriculture. *Int J Agric Sustain* 2010;**8**(4):219–36.
14. Kanianska R. Agriculture and its impact on land-use, environment, and ecosystems services. In: Almusaed A, editor. *Landscape ecology – the influences of land use and anthropogenic impacts of landscape creation*. Rijeka, Croatia: IntechOpen; 2016. p. 1–26.
15. Burguet M, Guzmán G, de Luna E, Taguas EV, Gómez JA. Evaluation of disruption of sediment connectivity and herbicide transport across a slope by grass strips using a magnetic iron oxide tracer. *Soil Res* 2018;**180**:268–81.

16. Seufert V, Ramankutty N, Foley JA. Comparing the yields of organic and conventional agriculture. *Nature* 2012;**485**:229–34.
17. De Ponti T, Rijk B, van Ittersum MK. The crop yield gap between organic and conventional agriculture. *Agric Syst* 2012;**108**:1–9.
18. Schrama M, de Haan JJ, Kroonen M, Verstegen H, Van der Putten WH. Crop yield gap and stability in organic and conventional farming systems. *Agric Ecosyst Environ* 2018;**256**:123–30.
19. Scialabba NEH, Műller-Leindenlauf M. Organic agriculture and climate change. *Renew Agric Food Syst* 2010;**25**(2):158–69.
20. Foley JA, DeFries RS, Asner GP, Barford C, Bonana G, Carpenter SR, Chapin FS. Global consequences of land use. *Science* 2005;**80**(309):570–4.
21. Gerard F, Petiti S, Smith G, Thomson A, Brown N, Manchester S, Wadsworth R, Bugar G, Halada L, Bezák P, Boltiziar M, De Badts E, Halabuk A, Mojses M, Petrovic F, Gregor M, Hazeu G, Mücher CA, Wachowicz M, Huitu H, Tuominen S, Köhler R, Olschofsky K, Ziese H, Kolar J, Sustera J, Luque S, Pino J, Pons X, Roda F, Roscher M, Feranec J. Land cover change in Europe between 1950 – 2000 determined employing aerial photography. *Prog Phys Geogr* 2010;**34**: 183–205.
22. Miao L, Zhu F, He B, Ferrat M, Liu Q, Cao X, Cui X. Synthesis of Chinas land use in the past 300 years. *Glob Planet Chang* 2013;**100**:224–33.
23. Slaughter RA. Welcome to the anthropocene. *Futures* 2011;**44**(2):19–26.
24. Persley GJ, Doyle JJ. *Biotechnology for developing – country agriculture: problem and opportunies. A 2020 vision for food, agriculture and the environment.* Washington, USA: International Food Policy Research Institute; 1999.
25. Ascough JC, Hoag DL, McMaster GS, Frasier W. Computer use and satisfaction by great plains producers. *Agron J* 2002;**94**:1263–9.
26. Bennet AB, Chi-Ham C, Barrows G, Sexton S, Zilberman D. Agricultural biotechnology: economics, environment, ethics and the future. *Annu Rev Environ Resour* 2013;**38**:249–79.
27. Pereira L, Wynberg R, Reis Y. Agroecology: the future of sustainable farming? *Environ Sci Policy Sustain Dev* 2018;**60**(4):4–17.
28. Blasi E, Passeri S, Galli FA. An ecological footprint approach to environmental - economic evaluation of farm results. *Agric Syst* 2016;**145**:76–82.
29. Clark M, Tilman D. Comparative analysis of environmental impacts of agricultural production systems, agricultural input efficiency, and food choice. *Environ Res Lett* 2017;**12**:064016.
30. Pereira P, Bogunovic I, Muñoz-Rojas M, Brevik EC. Soil ecosystem services, sustainability, valuation and management. *Curr Opin Environ Sci Health* 2018;**5**:7–13.
31. Zeder MA. The domestication of animals. *J Anthropol Res* 2012;**68**(2):161–90.
32. Grinin LE. Production revolutions and periodization of history: a comparative and theoretic-mathematical approach. *Soc Evol Hist* 2007;**6**(2):75–120.
33. Savolainen P. Domestication of dogs. In: Jensen P, editor. *The behavioural biology of dogs.* Wallingford, UK: CAB International; 2007. p. 21–37.
34. Rosen AM. *Civilizing climate: social responses to climate change in the ancient Near East.* Plymouth, United Kingdom: Altamira Press; 2007.
35. Zeder MA, Hesse B. The initial domestication of goats (*Capra hircus*) in the Zagros Mountains 10,000 years ago. *Science* 2000;**287**(5461):2254–7.
36. Ryder ML. *Sheep & man.* London, UK: Duckworth; 1983.
37. Chessa B, Pereira F, Arnaud F, Amorim A, Goyache F, Mainland I, Kao RR, Pemberton JM, Beraldi D, Stear MJ, Alberti A, Pittau M, Iannuzzi L, Banabazi MH, Kazwala RR, Zhang Y-p, Arranz JJ, Ali BA, Wang Z, Uzun M, Dione MM, Olsaker I, Holm L-E, Saarma U, Ahmad S, Marzanov N, Eythorsdottir E,

Holland MJ, Ajmone-Marsan P, Bruford MW, Kantanen J, Spencer TE, Palmarini M. Revealing the history of sheep domestication using retrovirus integrations. *Science* 2009;**324**(5926):532–6.
38. Pulido M, Barrena-González J, Badgery W, Rodrigo-Comino J, Cerdà A. Sustainable grazing. *Curr Opin Environ Sci Health* 2018;**5**:42–6.
39. Dhondt AA. Carrying capacity: a confusing concept. *Acta Oecol* 1988;**9**(4):337–46.
40. Rvd W. Do herbivores cause habitat degradation or vegetation state transition? Evidence from the tundra. *Oikos* 2006;**114**(1):177–86.
41. Shi X-M, Li XG, Li CT, Zhao Y, Shang ZH, Ma Q. Grazing exclusion decreases soil organic C storage at an alpine grassland of the Qinghai–Tibetan Plateau. *Ecol Eng* 2013;**57**:183–7.
42. Perevolotsky A, Seligman NG. Role of grazing in Mediterranean rangeland ecosystems. *Bioscience* 1998;**48**(12):1007–17.
43. Savory A. The Savory grazing method or holistic resource management. *Rangelands* 1983;**5**(4):155–9.
44. Voisin A, Bressou C, de Cuenca CL. *Productividad de la hierba.* Madrid: Tecnos; 1967.
45. Dangal SR, Tian H, Zhang B, Pan S, Lu C, Yang J. Methane emission from global livestock sector during 1890–2014: magnitude, trends and spatiotemporal patterns. *Glob Chang Biol* 2017;**23**(10):4147–61.
46. Oldeman LR. *Global extent of soil degradation. Bi-annual report 1991-1992/ISRIC.* Nairobi: ISRIC; 1992. p. 19–36.
47. Steinfeld H, Gerber P, Wassenaar T, Castel V, de Haan C. *Livestock's long shadow: environmental issues and options.* Rome, Italy: Food & Agriculture Organization; 2006.
48. Abu Hammad A, Tumeizi A. Land degradation: socioeconomic and environmental causes and consequences in the eastern Mediterranean. *Land Degrad Dev* 2012;**23**(3): 216–26.
49. Chiepe QM, Amantle IN, Roy Amos K. A review on grazing management systems and effects on savanna ecosystem dynamics. *Afr J Ecol Ecosyst* 2017;**4**(3):251–7.
50. Kruess A, Tscharntke T. Grazing intensity and the diversity of grasshoppers, butterflies, and trap-nesting bees and wasps. *Conserv Biol* 2002;**16**(6):1570–80.
51. Pulido M, Schnabel S, Lavado Contador JF, Lozano-Parra J, González F. The impact of heavy grazing on soil quality and pasture production in rangelands of SW Spain. *Land Degrad Dev* 2018;**29**(2):219–30.
52. Pineiro G, Paruelo JM, Oesterheld M, Jobbagy G. Pathways of grazing effects on soil organic carbon and nitrogen. *Rangel Ecol Manag* 2010;**63–1**:10–119.
53. Dlamini P, Chivenge P, Chaplot V. Overgrazing decreases soil organic carbon stocks the most under dry climates and low soil pH: a meta-analysis shows. *Agric Ecosyst Environ* 2016;**221**:258–69.
54. Resolution EU. *EU biodiversity strategy to 2020 (2011/2307(INI)).* 2012. European Parliament resolution, http://ec.europa.eu/environment/nature/biodiversity/comm2006/pdf/EP_resolution_april2012.pdf.
55. United Nations. *Transforming our world: the 2030 agenda for sustainable development.* United Nations; 2015. https://sustainabledevelopment.un.org/content/documents/21252030%20Agenda%20for%20Sustainable%20Development%20web.pdf.
56. Barbut M. Life on land – an introduction to goal 15. In: Trowbridge L, editor. *A Better World. Actions and commitments to the sustainable development goals.* United Kingdom: Gomme Press; 2018. p. 2–3.
57. Pereira P, Bogunovic I. Land degradation neutrality. How to reverse land degradation with conservation agriculture practices? In: Baban M, Rašić S, editors. *Agriculture in nature and environment protection.* Osijek, Croatia: Glas Slavonije d.d.; 2019. p. 36–43.
58. Wiesmeier M, Urbanski L, Hobley E, Lang B, von Lützow M, Marin-Spiotta E, van Wesemael Rabot E, Ließ M, Garcia-Franco N, Wollschläger U, Vogel HJ,

Kögel-Knabner I. Soil organic carbon storage as a key function of soils — a review of drivers and indicator at various scales. *Geoderma* 2019;**333**:149—62.
59. Birkás M, Antos G, Neményi M. *Environmentally-sound adaptable tillage*. Budapest: Akadémiai Kiadó; 2008.
60. Várallyay G. Soil resilience (Is soil a renewable natural resource?). *Cereal Res Commun* 2007;**35**(2):1277—80.
61. Matsi T. Liquid cattle manure application to soil and its effect on crop growth, yield, composition, and on soil properties. In: Whalen J, editor. *Soil fertility improvement and integrated nutrient management — a global perspective*. Rijeka, Croatia: In Tech — open science; 2012. p. 97—118.
62. Martin G, Moraine M, Ryschawy J, Magne MA, Asai M, Sarthou JP, Duru M, Therond O. Crop-livestock integration beyond the farm level: a review. *Agron Sustain Dev* 2016;**36**(53).
63. Birkas M, Dekemati I, Kende Z, Posa B. Review of soil tillage history and new challenges in Hungary. *Hung Geogr Bull* 2017;**66**(1):55—64.
64. Robertson GP, Gross KL, Hamilton SK, Landis DA, Schmidt TM, Snapp SS, Swinton SM. Farming for ecosystem services: an ecological approach to production agriculture. *Bioscience* 2014;**64**(5):404—15.
65. Montgomery DR. *Dirt — the erosion of civilizations*. London, England: University of California Press; 2012.
66. Bilandžija D, Zgorelec Ž, Kisić I. Influence of tillage practices and crop type on soil CO_2 emissions. *Sustainability* 2016;**8**(1):90.
67. Lal R. Accelerated soil erosion as a source of atmospheric CO_2. *Soil Res* 2019;**188**:35—40.
68. Sheibani S, Ahangar AG. Effect of tillage on soil biodiversity. *J Novel Appl Sci* 2013; **2**(8):273—81.
69. Carbonell-Bojollo R, González-Sánchez EJ, de Torres MRR, Ordóñez-Fernández R, Domínguez-Gimenez J, Basch G. Soil organic carbon fractions under conventional and no-till management in a long-term study in southern Spain. *Soil Res* 2015; **53**(2):113—24.
70. Kay BD, VandenBygaart AJ. Conservation tillage and depth stratification of porosity and soil organic matter. *Soil Res* 2002;**66**:107—18.
71. Glab T, Kulig B. Effect of mulch and tillage system on soil under wheat (*Triticum eastivum*). *Soil Res* 2008;**99**:169—78.
72. Franzluebbers A. Soil organic matter stratification ratio as an indicator of soil quality. *Soil Res* 2002;**66**(2):95—106.
73. Kisic I, Bogunovic I, Birkas M, Jurisic A, Spalevic V. The role of tillage and crops on a soil loss of an arable Stagnic Luvisol. *Arch Agron Soil Sci* 2017;**63**(3):403—13.
74. Sharma P, Abrol V, Sharma KR, Sharma N, Phogat VK, Vikas V. Impact of conservation tillage on soil organic carbon and physical properties a review. *Int J Bio-resource Stress Manag* 2016;**7**:151—61.
75. Lal R. Managing terrestrial carbon in a changing climate. In: Kapur S, Erşahin S, editors. *Soil security for ecosystem management*. Cham: Springer; 2014. p. 1—18.
76. Lu X, Lu X, Liao Y. Conservation tillage increases carbon sequestration of winter wheat-summer maize farmland on Loess Plateau in China. *PLoS One* 2018;**13**(9).
77. Lobb DA, Huffman T, Reicosky DC. Importance of information on tillage practices in the modelling of environmental processes and in the use of environmental indicators. *J Environ Manag* 2007;**82**(3):377—87.
78. Derpsch R, Friedrich T, Kassam A, Li H. Current status of adoption of no-till farming in the world and some of its main benefits. *Int J Agric Biol Eng* 2010;**3**(1):1—25.
79. Derpsch R, Franzluebbers AJ, Duiker SW, Reicosky DC, Koeller K, Friedrich T, Sturny WG, De Moraes Sa JD, Weis K. Why do we need standardize no-tillage research? *Soil Res* 2014;**137**:16—22.

80. Bot A, Benites J. *The importance of soil organic matter: key to drought-resistant soil and sustained food production*. Rome, Italy: Food & Agriculture Organisation; 2005.
81. Benbi DK, Nieder R, Isermann K. Soil organic matter dynamics. In: Benbi DK, Nieder R, editors. *Handbook of processes and modeling in the soil-plant system*. New York: CRC Press; 2003. p. 345–408.
82. Bastian F, Bouziri L, Nicolardot B, Ranjard L. Impact of wheat straw decomposition on successional patterns of soil microbial community structure. *Soil Biol Biochem* 2009; **41**(2):262–75.
83. Lal R. Land use and soil management effects on soil organic matter dynamics on Alfisols in Western Nigeria. In: Lal R, Kimble JM, Follett RF, Stewart BA, editors. *Soil processes and the carbon cycle*. Boca Raton: CRC Press; 2018. p. 109–26.
84. Fey MV, Mills AJ. Declining soil quality in South Africa: effects of land use on soil organic matter and surface crusting. *South Afr J Sci* 2003;**99**(9):429–36.
85. Matson PA, Parton WJ, Power A, Swift M. Agricultural intensification and ecosystem properties. *Science* 1997;**277**(5325):504–9.
86. Pulido-Fernández M, Schnabel S, Lavado-Contador JF, Miralles Mellado I, Ortega Pérez R. Soil organic matter of Iberian open woodland rangelands as influenced by vegetation cover and land management. *Catena* 2013;**109**:13–24.
87. Abdalla M, Hastings A, Chadwick D, Jones D, Evans C, Jones M, Rees R, Smith P. Critical review of the impacts of grazing intensity on soil organic carbon storage and other soil quality indicators in extensively managed grasslands. *Agric Ecosyst Environ* 2018;**253**:62–81.
88. Griffiths BS, Ritz K, Bardgett RD, Cook R, Christensen S, Ekelund F, Sørensen SJ, Bååth E, Bloem J, De Ruiter PC, Dolfing J, Nicolardot B. Ecosystem response of pasture soil communities to fumigation-induced microbial diversity reductions: an examination of the biodiversity–ecosystem function relationship. *Oikos* 2000;**90**(2): 279–94.
89. Moreno G, Obrador JJ, Cubera E, Dupraz C. Fine root distribution in dehesas of Central-Western Spain. *Plant Soil* 2005;**277**(1–2):153–62.
90. Pulido M, Schnabel S, Contador JFL, Lozano-Parra J, Gómez-Gutiérrez Á, Brevik EC, Cerdà A. Reduction of the frequency of herbaceous roots as an effect of soil compaction induced by heavy grazing in rangelands of SW Spain. *Catena* 2017;**158**:381–9.
91. Simón N, Montes F, Díaz-Pinés E, Benavides R, Roig S, Rubio A. Spatial distribution of the soil organic carbon pool in a Holm oak dehesa in Spain. *Plant Soil* 2013;**366**(1–2):537–49.
92. Joffre R, Rambal S. How tree cover influences the water balance of Mediterranean rangelands. *Ecology* 1993;**74**(2):570–82.
93. Ochoa-Hueso R, Eldridge DJ, Delgado-Baquerizo M, Soliveres S, Bowker MA, Gross N, Arredondo T. Soil fungal abundance and plant functional traits drive fertile island formation in global drylands. *J Ecol* 2018;**106**(1):242–53.
94. Milchunas DG, Lauenroth WK. Quantitative effects of grazing on vegetation and soils over a global range of environments. *Ecol Monogr* 1993;**63**(4):327–66.
95. Baker H. Patterns of plant invasion in North America. In: Mooney HA, Drake JA, editors. *Ecology of biological invasions of North America and Hawaii*. New York: Springer; 1986. p. 44–57.
96. Eichhorn M, Paris P, Herzog F, Incoll L, Liagre F, Mantzanas K, Mayus M, Moreno G, Papanastasis V, Pilbeam D. Silvoarable systems in Europe–past, present and future prospects. *Agrofor Syst* 2006;**67**(1):29–50.
97. Fuhlendorf SD, Engle DM. Restoring heterogeneity on rangelands: ecosystem management based on evolutionary grazing patterns: we propose a paradigm that enhances heterogeneity instead of homogeneity to promote biological diversity and wildlife habitat on rangelands grazed by livestock. *Bioscience* 2001;**51**(8):625–32.

98. Paustian K, Collins HP, Paul EA. Management controls on soil carbon. In: Paul EA, Paustian KH, Elliot ET, Vernon Cole C, editors. *Soil organic matter in temperate agroecosystems. Long term experiments in North America.* Boca Raton, FL, USA: CRC Press; 1997. p. 15–49.
99. Carter MR. Soil quality for sustainable land management. *Agron J* 2002;**94**(1):38–47.
100. Bünemann EK, Bongiorno G, Bai Z, Creamer RE, De Deyn G, de Goede R, Pulleman M. Soil quality—A critical review. *Soil Biol Biochem* 2018;**120**:105–25.
101. Nawaz MF, Bourrie G, Trolard F. Soil compaction impact and modelling. A review. *Agron Sustain Dev* 2013;**33**(2):291–309.
102. Bogunovic I, Bilandzija D, Andabaka Z, Stupic D, Comino JR, Cacic M, Pereira P. Soil compaction under different management practices in a Croatian vineyard. *Arab J Geosci* 2017;**10**(15):340.
103. Bronick CJ, Lal R. Soil structure and management: a review. *Geoderma* 2005;**124**(1–2):3–22.
104. Vaezi AR, Ahmadi M, Cerdà A. Contribution of raindrop impact to the change of soil physical properties and water erosion under semi-arid rainfalls. *Sci Total Environ* 2017; **583**:382–92.
105. Di Prima S, Rodrigo-Comino J, Novara A, Iovino M, Pirastru M, Keesstra S, Cerdà A. Soil physical quality of citrus orchards under tillage, herbicide, and organic managements. *Pedosphere* 2018;**28**(3):463–77.
106. Goulding KWT. Soil acidification and the importance of liming agricultural soils with particular reference to the United Kingdom. *Soil Use Manag* 2016;**32**(3):390–9.
107. Xin X, Zhang J, Zhu A, Zhang C. Effects of long-term (23 years) mineral fertilizer and compost application on physical properties of fluvo-aquic soil in the North China Plain. *Soil Res* 2016;**156**:166–72.
108. Six J, Elliott ET, Paustian K. Soil structure and soil organic matter II. A normalized stability index and the effect of mineralogy. *Soil Sci Soc Am J* 2000;**64**(3):1042–9.
109. Jenny H. *Factors of soil formation: a system of quantitative pedology.* New York, USA: Dower Publications; 1941.
110. Zech W, Senesi N, Guggenberger G, Kaiser K, Lehmann J, Miano TM, Schroth G. Factors controlling humification and mineralization of soil organic matter in the tropics. *Geoderma* 1997;**79**(1–4):117–61.
111. Six J, Conant RT, Paul EA, Paustian K. Stabilization mechanisms of soil organic matter: implications for C-saturation of soils. *Plant Soil* 2002;**241**(2):155–76.
112. Zhang GS, Chan KY, Oates A, Heenan DP, Huang GB. Relationship between soil structure and runoff/soil loss after 24 years of conservation tillage. *Soil Res* 2007; **92**(1–2):122–8.
113. Birkás M, Jug D, Kisic I. *Book of soil tillage.* Gödöllő: Szent István Egyetemi Kiadó; 2014.
114. Busari MA, Kukal SS, Kaur A, Bhatt R, Dulazi AA. Conservation tillage impacts on soil, crop and the environment. *Int Soil Water Conserv Res* 2015;**3**(2):119–29.
115. Tormena CA, Karlen DL, Logsdon S, Cherubin MR. Visual soil structure effects of tillage and corn stover harvest in Iowa. *Soil Sci Soc Am J* 2016;**80**(3):720–6.
116. Bogunovic I, Pereira P, Kisic I, Sajko K, Sraka M. Tillage management impacts on soil compaction, erosion and crop yield in Stagnosols (Croatia). *Catena* 2018;**160**:376–84.
117. Van Zanten BT, Verburg PH, Espinosa M, Gomez-y-Paloma S, Galimberti G, Kantelhardt J, Raggi M. European agricultural landscapes, common agricultural policy and ecosystem services: a review. *Agron Sustain Dev* 2014;**34**(2):309–25.
118. Blevins RL, Lal R, Doran JW, Langdale GW, Frye WW. Conservation tillage for erosion control and soil quality. In: Pierce FJ, Frye WW, editors. *Advances in soil and water conservation.* New York: Routledge; 2018. p. 51–68.

119. Tisdall JM, Oades J. Organic matter and water-stable aggregates in soils. *J Soil Sci* 1982; **33**(2):141–63.
120. Lal R. Restoring soil quality to mitigate soil degradation. *Sustainability* 2015;**7**(5): 5875–95.
121. Pagliai M, Vignozzi N, Pellegrini S. Soil structure and the effect of management practices. *Soil Res* 2004;**79**(2):131–43.
122. Bogunovic I, Trevisani S, Pereira P, Vukadinovic V. Mapping soil organic matter in the Baranja region (Croatia): geological and anthropic forcing parameters. *Sci Total Environ* 2018;**643**:335–45.
123. Bogunovic I, Kisic I. Compaction of a clay loam soil in Pannonian region of Croatia under different tillage systems. *J Agric Sci Technol* 2017;**19**(2):475–86.
124. Schneider F, Don A, Hennings I, Schmittmann O, Seidel SJ. The effect of deep tillage on crop yield—What do we really know? *Soil Res* 2017;**174**:193–204.
125. Six JAET, Elliott ET, Paustian K. Soil macroaggregate turnover and microaggregate formation: a mechanism for C sequestration under no-tillage agriculture. *Soil Biol Biochem* 2000;**32**:2099–103.
126. Carter MR. *Conservation tillage in temperate agroecosystems*. Boca Raton, USA: CRC Press; 2017.
127. Alvarez R, Steinbach HS. A review of the effects of tillage systems on some soil physical properties, water content, nitrate availability and crops yield in the Argentine Pampas. *Soil Res* 2009;**104**(1):1–15.
128. Crittenden SJ, Poot N, Heinen M, Van Balen DJM, Pulleman MM. Soil physical quality in contrasting tillage systems in organic and conventional farming. *Soil Res* 2015; **154**:136–44.
129. Wang Y, Li C, Tu C, Hoyt GD, DeForest JL, Hu S. Long-term no-tillage and organic input management enhanced the diversity and stability of soil microbial community. *Sci Total Environ* 2017;**609**:341–7.
130. Aksakal EL, Sari S, Angin I. Effects of vermicompost application on soil aggregation and certain physical properties. *Land Degrad Dev* 2016;**27**(4):983–95.
131. Mujdeci M, Isildar AA, Uygur V, Alaboz P, Unlu H, Senol H. Cooperative effects of field traffic and organic matter treatments on some compaction-related soil properties. *Solid Earth* 2017;**8**(1):189.
132. Whalen JK, Chang C. Macroaggregate characteristics in cultivated soils after 25 annual manure applications. *Soil Sci Soc Am J* 2002;**66**(5):1637–47.
133. Hati K, Bandyoopadhay K. Fertilizers (mineral, organic), effect on soil physical properties. In: Gliński J, Horabik J, Lipiec J, editors. *Encyclopedia of agrophysics*. Dordrecht: Springer; 2011. p. 296–9.
134. Sadeghi SHR, Gholami L, Sharifi E, Khaledi Darvishan A, Homaee M. Scale effect on runoff and soil loss control using rice straw mulch under laboratory conditions. *Solid Earth* 2015;**6**(1):1–8.
135. Diacono M, Montemurro F. Long-term effects of organic amendments on soil fertility. *Agron Sustain Dev* 2010;**30**:401–22.
136. Agegnehu G, Srivastava AK, Bird MI. The role of biochar and biochar-compost in improving soil quality and crop performance: a review. *Appl Soil Ecol* 2017;**119**: 156–70.
137. De León-González F, Hernández-Serrano MM, Etchevers JD, Payan-Zelaya F, Ordaz-Chaparro V. Short-term compost effect on macroaggregation in a sandy soil under low rainfall in the valley of Mexico. *Soil Res* 2000;**56**(3–4):213–7.
138. Haynes RJ, Naidu R. Influence of lime, fertilizer and manure applications on soil organic matter content and soil physical conditions: a review. *Nutrient Cycl Agroecosyst* 1998;**51**:123–37.

139. Munkholm LJ, Schjønning P, Debosz K, Jensen HE, Christensen BT. Aggregate strength and mechanical behaviour of a sandy loam soil under long-term fertilization treatments. *Eur J Soil Sci* 2002;**53**(1):129–37.
140. Stagnari F, Perpertuini G, Tofalo R, Campanelli G, Leteo F, Della Vella U, Schirone M, Suzzi G, Pisante M. Long-term impact of farm management and crops on soil microorganisms assessed by combined DGGE and PLFA analyses. *Front Microbiol* 2014;**5**:644.
141. Guo LB, Gifford RM. Soil carbon stocks and land use change: a meta analysis. *Glob Chang Biol* 2002;**8**(4):345–60.
142. Raimbault BA, Vyn TJ. Crop rotation and tillage effects on corn growth and soil structural stability. *Agron J* 1991;**83**(6):979–85.
143. Kong AY, Six J, Bryant DC, Denison RF, Van Kessel C. The relationship between carbon input, aggregation, and soil organic carbon stabilization in sustainable cropping systems. *Soil Sci Soc Am J* 2005;**69**(4):1078–85.
144. Yeboah S, Zhang R, Cai L, Li L, Xie J, Luo Z, Liu J, Wu J. Tillage effect on soil organic carbon, microbial biomass carbon and crop yield in spring wheat-field pea rotation. *Plant Soil Environ* 2016;**62**(6):279–85.
145. Castro Filho CD, Lourenço A, Guimarães MDF, Fonseca ICB. Aggregate stability under different soil management systems in a red latosol in the state of Parana, Brazil. *Soil Res* 2002;**65**(1):45–51.
146. Greenwood KL, McKenzie BM. Grazing effects on soil physical properties and the consequences for pastures: a review. *Aust J Exp Agric* 2001;**41**(8):1231–50.
147. Proffitt A, Jarvis R, Bendotti S. The impact of sheep trampling and stocking rate on the physical properties of a red duplex soil with two initially different structures. *Aust J Agric Res* 1995;**46**(4):733–47.
148. Hamza MA, Anderson WK. Soil compaction in cropping systems: a review of the nature, causes and possible solutions. *Soil Res* 2005;**82**(2):121–45.
149. Kondo M, Junior D. Soil compressibility of three Latosols (oxisols) as a function of moisture and use. *Rev Bras Ciênc Solo* 1999;**23**(2):211–8.
150. Bilotta GS, Brazier RE, Haygarth PM. The impacts of grazing animals on the quality of soils, vegetation, and surface waters in intensively managed grasslands. *Adv Agron* 2007; **94**:237–80.
151. Evans R. Soil erosion in the UK initiated by grazing animals. *Geography* 1997;**17**(2): 127–41.
152. Cerdà A. Aggregate stability against water forces under different climates on agriculture land and scrubland in southern Bolivia. *Soil Res* 2000;**57**(3):159–66.
153. Grace P, Oades J, Keith H, Hancock T. Trends in wheat yields and soil organic carbon in the permanent rotation trial at the waite agricultural research Institute, south Australia. *Aust J Exp Agric* 1995;**35**(7):857–64.
154. Nearing MA, Xie Y, Liu B, Ye Y. Natural and anthropogenic rates of soil erosion. *Int Soil Water Conserv Res* 2017;**5**:77–84.
155. Lindstrom MJ, Lobb DA, Schumacher TE. Tillage erosion: an overview. *Ann Arid Zone* 2001;**40**(3):337–49.
156. Li L, Du S, Wu L, Liu G. An overview of soil loss tolerance. *Catena* 2009;**78**:93–9.
157. Pimentel D, Burgess M. Soil erosion threatens food production. *Agriculture* 2013;**3**: 443–63.
158. Blanco-Canqui H, Lal R. *Principles of soil conservation and management*. Dordrecht: Springer; 2010.
159. Boardman J, Poesen J. *Soil erosion in Europe*. Chichester, United Kingdom: John Wiley & Sons; 2006.
160. Pimentel D. Soil erosion: a food and environmental threat. *Environ Dev Sustain* 2006;**8**: 119–37.

161. Kisic I, Bogunovic I, Bilandzija D. The influence of tillage and crops on particle size distribution of water eroded soil sediment on stagnosol. *Soil Water Res* 2017;**12**(3): 170–6.
162. Kisic I, Bogunovic I, Zgorelec Z, Bilandzija D. Effects of soil erosion by water under different tillage treatments on distribution of soil chemical parameters. *Soil Water Res* 2018;**13**(1):36–43.
163. Wendt RC, Burwell RE. Runoff and soil losses for conventional, reduced, and no-till corn. *J Soil Water Conserv* 1985;**40**:450–4.
164. Borrelli P, Paustian K, Panagos P, Jones A, Schütt B, Lugato E. Effect of good agricultural and environmental conditions on erosion and soil organic carbon balance: a national case study. *Land Use Policy* 2016;**50**:408–21.
165. Prosdocimi M, Jordán A, Tarolli P, Keesstra S, Novara A, Cerdà A. The immediate effectiveness of barley straw mulch in reducing soil erodibility and surface runoff generation in Mediterranean vineyards. *Sci Total Environ* 2016;**547**:323–30.
166. Zuzel JF, Pikul Jr JL. Effects of straw mulch on runoff and erosion from small agricultural plots in northeastern Oregon. *Soil Sci* 1993;**156**:111–7.
167. Derpsch R, Sidiras N, Roth CH. Results of studies made from 1977 to 1984 to control erosion by cover crops and no-tillage techniques in Parana, Brazil. *Soil Res* 1986;**8**: 253–63.
168. Borrelli P, Ballabio C, Panagos P, Montanarella L. Wind erosion susceptibility of European soils. *Geoderma* 2014;**232–234**:471–8.
169. Chepil WS. Conversion of relative field erodibility to annual soil loos by wind. *Soil Sci Soc Am* 1960;**24**:143–5.
170. Zamani S, Mahmoodabadi M. Effect of particle-size distribution on wind erosion rate and soil erodibility. *Arch Agron Soil Sci* 2012;**59**(12):1743–53.
171. Zobeck TM, Baddock M, Van Pelt RS, Tatarko J, Acosta-Martinez V. Soil property effects on wind erosion of organic soils. *Aeolian Res* 2013;**10**:43–51.
172. Washington R, Todd M, Lizcano G, Tegen I, Flamant C, Koren I, Ginoux P, Engelstaedter S, Bristow C, Zender C. Links between topography, wind, deflation, lakes and dust: the case of the Bodélé Depression, Chad. *Geophys Res Lett* 2006; **33**(9):1–4.
173. Lyles L, Tatarko J. Wind erosion effects on soil texture and organic matter. *J Soil Water Conserv* 1986;**41**(3):191–3.
174. Sterk G. Causes, consequences and control of wind erosion in Sahelian Africa: a review. *Land Degrad Dev* 2003;**14**:95–108.
175. Uri ND. Agriculture and environment: the problem of soil erosion. *J Sustain Agric* 2000;**16**(4):71–94.
176. Riksen MJPM, de Graaff J. On-site and off-site effects of wind erosion on European light soils. *Land Degrad Dev* 2001;**12**:1–11.
177. Tegen I, lacis AA, Fung IY. The influence of mineral aerosol from disturbed soils on global radiation budget. *Nature* 1996;**380**:419–22.
178. Polymenakou PN, Mandalakis M, Stephanou EG, Tselepides A. Particle size distribution of airborne microorganisms and pathogens during an intense african dust event in the eastern mediterranean. *Environ Health Perspect* 2008;**116**(3):292–6.
179. Panagos P, Borrelli P, Poesen J. Soil loss due to crop harvesting in the European Union: a first estimation of an underrated geomorphic process. *Sci Total Environ* 2019;**664**:487–98.
180. Mwango SB, Msanya BM, Mkatwa PW, Kimaro DN, Deckers S, Poesen J, Sanga E. Soil loss due to crop harvesting in Usambara mountains Tanzania. The case of carrot, onion and potato. *Int J Plant Soil Sci* 2015;**4**(1):18–28.
181. Tugrul KM, Icöz E, Perendeci NA. Determination of soil loss by sugar beet harvesting. *Soil Res* 2012;**123**:71–7.

182. Parlak M, Blanco-Canqui H. Soil losses due to potato harvesting: a case study in western Turkey. *Soil Use Manag* 2015;**31**:525–7.
183. Sumithra R, Thushyanthy M, Srivaratharasan T. Assessment of soil loss and nutrient depletion due to cassava harvesting: a case study from low input traditional agriculture. *Int Soil Water Conserv Res* 2013;**1**(2):72–9.
184. Dada POO, Adeyanju OR, Adeosun OJ, Adewumi JK. Effects of soil physical loss due to manual yam harvesting under a sandy loam environment. *Int Soil Water Conserv Res* 2016;**4**:121–5.
185. Poesen J, Verstraeten G, Soenens R, Seynaeve L. Soil losses due to harvesting of chicory roots and sugar beet: an underated geomorphic process. *Catena* 2001;**43**(1): 35–47.
186. Jurisic A, Kisic I, Basic F, Zgorelec Z, Matotek S. Soil losses and soil degradation processes caused by harvest of sugar beet. *Novenytermeles* 2011;**60**:255–8.
187. Ruysschaert G, Poesen J, Verstraeten G, Govers G. Soil loss due to harvesting of various crop types in contrasting agro-ecological environments. *Agric Ecosyst Environ* 2007;**120**:153–65.
188. Schnabel S. *Soil erosion and runoff production in a small watershed under silvo-pastoral landuse (dehesas) in Extremadura, Spain.* Logroño, Spain: Geoforma Ediciones; 1997.
189. Gómez Gutiérrez Á, Schnabel S, Felicísimo ÁM. Modelling the occurrence of gullies in rangelands of southwest Spain. *Earth Surf Process Landforms* 2009;**34**(14):1894–902.
190. Schnabel S, González F, Murillo Villanova M, Moreno V. Different techniques of pasture improvement and soil erosion in a wooded rangeland in SW Spain. Methodology and preliminary results. In: Conacher A, editor. *Land degradation.* Dordrecht, The Netherlands: Kluwer Academic Publishers; 2001. p. 241–57.
191. Schnabel S, Gómez Gutiérrez Á, Lavado Contador JF. Grazing and soil erosion in dehesas of SW Spain. In: Romero Díaz A, Belmonte Serrato F, Alonso Sarria F, López Bermúdez F, editors. *Advances in studies on desertification.* Murcia: Editum; 2009. p. 725–8.
192. Gómez JA, Campos MG, Castillo F, Vanwalleghem T, Giraldez JV. Soil erosion control, plant diversity, and arthropod communities under heterogeneous cover crops in an olive orchard. *Environ Sci Pollut Res* 2018;**25**:977–89.
193. Sala M. Cobertura vegetal y respuesta hidrológica. Ejemplo de las Cordilleras Costeras Catalanas. In: Campesino Fernández AJ, Velasco Bernardo C, editors. *España-Portugal: Ordenación Territorial del Sureste Comunitario.* Cáceres, Spain: Universidad de Extremadura; 1996. p. 177–8.
194. Zhao HL, Zhao XY, Zhou RL, Zhang TH, Drake S. Desertification processes due to heavy grazing in sandy rangeland, Inner Mongolia. *J Arid Environ* 2005;**62**(2):309–19.
195. Hoffmann C, Funk R, Li Y, Sommer M. Effect of grazing on wind driven carbon and nitrogen ratios in the grasslands of Inner Mongolia. *Catena* 2008;**75**(2):182–90.
196. Al-Dousari AM, Misak R, Shahid S. Soil compaction and sealing in Al-Salmi area, Western Kuwait. *Land Degrad Dev* 2000;**11**:401–18.
197. Botta GF, Tolon-Becerra A, Lastra-Bravo X, Tourn M. Tillage and traffic effects (planters and tractors) on soil compaction and soybean (*Glycine max* L.) yields in Argentinean pampas. *Soil Res* 2010;**110**(1):167–74.
198. Dexter AR. Soil physical quality: part I. Theory, effects of soil texture, density, and organic matter, and effects on root growth. *Geoderma* 2004;**120**:201–14.
199. Teepe R, Brumme R, Beese F, Ludig B. Nitrous oxide emission and methane consumption following compaction of forest soils. *Soil Sci Soc Am J* 2004;**68**:605-611.
200. Gaertig T, Schack-Kirchner H, Hildebrand EE, von Wilpert K. The impact of soil aeration on oak decline in southwestern Germany. *For Ecol Manag* 2002;**159**:15–25.
201. Woodward CL. Soil compaction and topsoil removal effects on soil properties and seedling growth in Amazonian Ecuador. *For Ecol Manag* 1996;**82**:197-209.

202. Battigelli JP, Spence JR, Langor DW, Berch SM. Short-term impact of forest soil compaction and organic matter removal on soil mesofauna density and oribatid mite diversity. *Can J Res* 2004;**34**:1136-1149.
203. Boström U. The effect of soil compaction on earthworms (*Lumbricidae*) in a heavy clay soil. *Swed J Agric Res* 1986;**16**:137-141.
204. Birkás M, Szalai T, Gyuricza C, Gecse M, Bordas K. Effects of disk tillage on soil condition, crop yield and weed infestation. *Rostl Vyroba* 2002;**48**(2):20–6.
205. Birkás M, Jolánkai M, Gyuricza C, Percze A. Tillage effects on compaction, earthworms and other soil quality indicators in Hungary. *Soil Res* 2004;**78**(2):185–96.
206. Bašić F, Herceg N. *Principles of agronomy*. Zagreb: Synopsis; 2010 (in Croatian).
207. Neve S, Hofman G. Influence of soil compaction on carbon and nitrogen mineralization of soil organic matter and crop residues. *Biol Fertil Soils* 2000;**30**:544–9.
208. Conlin TSS, Driessche R. Response of soil CO_2 and O_2 concentrations to forest soil compaction at the long-term soil productivity sites in central British Columbia. *Can J Soil Sci* 2000;**80**:625–32.
209. Agrawal RP. Water and nutrient management in sandy soils by compaction. *Soil Res* 1991;**19**:121-130.
210. Brais S. Persistence of soil compaction and effects on seedling growth in Northwestern Quebec. *Soil Sci Soc Am J* 2001;**65**:1263-1271.
211. Flowers M, Lal R. Axle load and tillage effect on soil physical properties and soybean grain yield on a mollic ochraqualf in northwest Ohio. *Soil Res* 1998;**48**:21–35.
212. Carder J, Grasby J. *A framework for regional soil conservation treatments in the medium and low rainfall agricultural district*. Research Report 1/86. Western Australia: Department of Agriculture; 1986. p. 120.
213. Van den Akker JJH, Canarache A. Two European concerted actions on subsoil compaction. *Z Kult Landentwickl* 2001;**42**(1):15–22.
214. Osman KT. *Soil degradation, conservation and remediation*. The Netherlands: Springer; 2014.
215. Duttmann R, Schwanebeck M, Nolde M, Horn R. Predicting soil compaction risks related to field traffic during silage maize harvest. *Soil Sci Soc Am J* 2014;**78**(2):408–21.
216. Hakansson I, Petelkau H. Benefits of limited axle load. In: Soane BD, Van Ouwerkerk C, editors. *Soil compaction in crop production. Developments in agricultural engineering*. The Netherlands: Elsevier; 1994. p. 479–500.
217. Alakukku L, Weisskopf P, Chamen WCT, Tijink FGJ, Van der Linden JP, Pires S, Spoor G. Prevention strategies for field traffic-induced subsoil compaction: a review: part 1. Machine/soil interactions. *Soil Res* 2003;**73**(1–2):145–60.
218. Butorac A. *General agronomy*. Zagreb: Školska knjiga; 1999 (in Croatian).
219. Håkansson I. *Machinery-induced compaction of arable soils*. Uppsala, Sweden: Swedish University of Agricultural Sciences; 2005.
220. Cambi M, Certini G, Neri F, Marchi E. The impact of heavy traffic on forest soils: a review. *Ecol Manag* 2015;**338**:124–38.
221. Pagliai M, Marsili A, Servadio P, Vignozzi N, Pellegrini S. Changes in some physical properties of a clay soil in Central Italy following the passage of rubber tracked and wheeled tractors of medium power. *Soil Res* 2003;**73**(1–2):119–29.
222. Botta GF, Jorajuria D, Rosatto H, Ferrero C. Light tractor traffic frequency on soil compaction in the Rolling Pampa region of Argentina. *Soil Res* 2006;**86**(1):9–14.
223. Ampoorter E, De Schrijver A, Van Nevel L, Hermy M, Verheyen K. Impact of mechanized harvesting on compaction of sandy and clayey forest soils: results of a meta-analysis. *Ann For Sci* 2012;**69**(5):533–42.
224. Brais S, Camiré C. Soil compaction induced by careful logging in the claybelt region of northwestern Quebec (Canada). *Can J Soil Sci* 1998;**78**:197-206.

225. Ampoorter E, Goris R, Cornelis WM, Verheyen K. Impact of mechanized logging on compaction status of sandy forest soils. *Ecol Manag* 2007;**241**(1–3):162–74.
226. Ellies S, Smith R, Pröschle T. Effect of moisture and transit frequency on stress distribution on different soils. *Agro Sur* 2000;**28**(2):60–8.
227. Chan KY, Oates A, Swan AD, Hayes RC, Dear BS, Peoples MB. Agronomic consequences of tractor wheel compaction on a clay soil. *Soil Res* 2006;**89**:13–21.
228. Blanco-Sepúlveda R. La relación entre la densidad aparente y la resistencia mecánica como indicadores de la compactación del suelo. *Agrociencia* 2009;**43**:231–9.
229. Awadhwall NK, Thierstein GE. Soil crust and its impact on crop establishment: a review. *Soil Res* 1985;**5**(3):289–302.
230. Li B, Gao J, Wong X, Ma L, Cui Q, Vest M. Effects of biological soil crusts on water infiltration and evaporation Yanchi Ningxia, Maowusu Desert, China. *Int J Sediment Res* 2016;**31**(4):311–23.
231. Šimansky V, Pollakova N, Halmo S. Soil crust in agricultural land. *Acta Fytotechn Zootechn* 2004;**17**(4):109–14.
232. Castilho SCDP, Cooper M, Juhász CEP. Influence of crust formation under natural rain on physical attributes of soils with different textures. *Rev Bras Ciênc Solo* 2011; **35**(6):1893–905.
233. Bu C, Wu S, Yang K. Effects of physical soil crusts on infiltration and splash erosion in three typical Chinese soils. *Int J Sediment Res* 2014;**29**(4):491–501.
234. Lal R, Shukla MK. *Principles of soil physics*. New York: Marcel Dekker; 2004.
235. Akasheh OZ, Abu-Awwad AM. Irrigation and soil surface management in arid soils with surface crust. *J Arid Environ* 1997;**37**(2):243–50.
236. Uson A, Poch RM. Effects of tillage and management practices on soil crust morphology under a Mediterranean environment. *Soil Res* 2000;**54**:191–6.
237. Song G, Li X, Hui R. Biological soil crusts determine the germination and growth of two exotic plants. *Ecol Evol* 2017;**7**(9):441–9450.
238. Feng W, Zhang Y, Jia X, Wu B, Zha T, Qin S, Wang B, Shao C, Liu J, Fa K. Impact of environmental factors and biological soil crust types on soil respiration in a desert ecosystem. *PLoS One* 2014;**9**(7).
239. Le Bissonnais YL. Aggregate stability and assessment of soil crustability and erodibility: i. Theory and methodology. *Eur J Soil Sci* 1996;**47**(4):425–37.
240. Anderson DC, Harper K, Rushforth S. Recovery of cryptogamic soil crusts from grazing on Utah winter ranges. *J Range Manage Arch* 1982;**35**(3):355–9.
241. Belnap J, Gillette DA. Vulnerability of desert biological soil crusts to wind erosion: the influences of crust development, soil texture, and disturbance. *J Arid Environ* 1998; **39**(2):133–42.
242. Peter G, Funk FA. Montículos fitogéneticos en el Monte Rionegrino: su relación con incendios y pastoreo. *BioScriba* 2010;**3**(1):47–54.
243. Hiernaux P, Bielders CL, Valentin C, Bationo A, Fernández-Rivera S. Effects of livestock grazing on physical and chemical properties of sandy soils in Sahelian rangelands. *J Arid Environ* 1999;**41**(3):231–45.
244. Anderson DC, Harper KT, Holmgren RC. Factors influencing development of cryptogamic soil crusts in Utah deserts. *J Range Manage Arch* 1982;**35**(2):180–5.
245. Eldridge D, Greene R. Microbiotic soil crusts-a review of their roles in soil and ecological processes in the rangelands of Australia. *Soil Res* 1994;**32**(3):389–415.
246. Graetz RD, Tongway DJ. Influence of grazing management on vegetation, soil structure and nutrient distribution and the infiltration of applied rainfall in a semi-arid chenopod shrubland. *Aust J Ecol* 1986;**11**(4):347–60.
247. Muscha JM, Hild AL. Biological soil crusts in grazed and ungrazed Wyoming sagebrush steppe. *J Arid Environ* 2006;**67**(2):195–207.

248. Aguilar AJ, Huber-Sannwald E, Belnap J, Smart D, Moreno JA. Biological soil crusts exhibit a dynamic response to seasonal rain and release from grazing with implications for soil stability. *J Arid Environ* 2009;**73**(12):1158–69.
249. Oldeman LR. *An international methodology for an assessment of soil degradation, land georeferenced soils and terrain database.* RAPA Publication (FAO); 1994.
250. Vukadinović V, Vukadinović V. *Plant nutrition.* Osijek: Faculty of Agriculture Osijek; 2011 (in Croatian).
251. Foy CD. Physiological effects of hydrogen, aluminum, and manganese toxicities in acid soil. In: Adams F, editor. *Soil acidity and liming.* Madison, USA: Agron. Soc. Am; 1984. p. 57–97.
252. Lauber CL, Hamady M, Knight R, Fierer N. Pyrosequencing-based assessment of soil pH as a predictor of soil bacterial community structure at the continental scale. *Appl Environ Microbiol* 2009;**75**(15):5111–20.
253. Zebarth BJ, Forge TA, Goyer C, Brin LD. Effect of soil acidification on nitrification in soil. *Can J Soil Sci* 2015;**95**(4):359–63.
254. Stempfhuber B, Richter-Heitmann T, Bienek L, Schöning I, Schrumpf M, Friedrich M, Schloter M. Soil pH and plant diversity drive co-occurrence patterns of ammonia and nitrite oxidizer in soils from forest ecosystems. *Biol Fertil Soils* 2017; **53**(6):691–700.
255. Zgorelec Ž, Pehnec G, Bašić F, Kisić I, Mesić M, Žužul S, Jurišić A, Šestak I, Vađić V, Čačković M. Sulphur cycling between terrestrial agroecosystem and atmosphere. *Arh Hig Rada Toksikol* 2012;**63**(3):301–10.
256. Mesić M, Bogunović I. Importance of crop fertilization. In: Bogunović I, editor. *Sustainable soil management measures in organic agriculture for climate conditions of Mediterranean Croatia.* Zagreb: University of Zagreb Faculty of Agriculture; 2018.
257. Johnston AE, Goulding KWT, Poulton PR. Soil acidification during more than 100 years under permanent grassland and woodland at Rothamsted. *Soil Use Manag* 1986;**2**(1):3–10.
258. Tisdale SL, Nelson WL, Beaton JD. *Soil fertility and fertilizers.* New York: MacMillan; 1985.
259. Barak P, Jobe BO, Krueger AR, Peterson LA, Laird DA. Effects of long-term soil acidification due to nitrogen fertilizer inputs in Wisconsin. *Plant Soil* 1997;**197**(1):61–9.
260. Gavi F, Basta NT, Raun WR. Wheat grain cadmium as affected by long-term fertilization and soil acidity. *J Environ Qual* 1997;**26**(1):265–72.
261. Pierre WH, Webb JR, Shrader WD. Quantitative effects of nitrogen fertilizer on the development and downward movement of soil acidity in relation to level of fertilization and crop removal in a continuous corn cropping system. *Agron J* 1971;**63**(2):291–7.
262. Robarge W. Acidity. In: Chesworth W, editor. *Encyclopedia of soil science.* Netherlands: Springer; 2008. p. 10–21.
263. Bolan NS, Hedley MJ. Role of carbon, nitrogen and sulfur cycles in soil acidification. In: Rengel Z, editor. *Handbook of soil acidity.* New York: Marcel Dekker; 2003. p. 29–52.
264. Robson A. *Soil acidity and plant growth.* The Netherland: Elsevier; 2012.
265. Mesic M, Zgorelec Z, Percin A, Bogunovic I, Bilandzija D. Secondary macronutrients: advances in understanding calcium cycling in soils, uptake/use by plants and ways of optimising calcium use efficiency in crop production. In: Zengel R, editor. *Achieving sustainable crop nutrition.* Cambridge, United Kingdom: Burleigh Dodds Science Publishing Limited; 2019 (in press).
266. Kisić I, Bašić F, Mesić M, Butorac A, Vaic Ž. The effect of liming and fertilization on yields of maize and winter wheat. *Agric Conspectus Sci* 2004;**69**(2–3):51–7.

267. Six J, Bossuyt H, Degryze S, Denef K. A history of research on the link between (micro) aggregates, soil biota, and soil organic matter dynamics. *Soil Res* 2004;**79**(1):7–31.
268. Martins AP, Andrade Costa SEVG, Anghinoni I, Kunrath TR, Balerini F, Cecagno D, Carvalho PCF. Soil acidification and basic cation use efficiency in an integrated no-till crop–livestock system under different grazing intensities. *Agric Ecosyst Environ* 2014; **195**:18–28.
269. Dorrough J, Yen A, Turner V, Clark SG, Crosthwaite J, Hirth JR. Livestock grazing management and biodiversity conservation in Australian temperate grassy landscapes. *Aust J Agric Res* 2004;**55**(3):279–95.
270. Peters GM, Wiedemann S, Rowley HV, Tucker R, Feitz AJ, Schulz M. Assessing agricultural soil acidification and nutrient management in life cycle assessment. *Int J Life Cycle Assess* 2011;**16**(5):431.
271. Bolan N, Hedley M, White R. Processes of soil acidification during nitrogen cycling with emphasis on legume based pastures. *Plant Soil* 1991;**134**(1):53–63.
272. Bell LW, Moore AD, Kirkegaard JA. Evolution in crop–livestock integration systems that improve farm productivity and environmental performance in Australia. *Eur J Agron* 2014;**57**:10–20.
273. Martins AP, Anghinoni I, Costa SEVGA, Carlos FS, Nichel GH, Silva RAP, Carvalho PCF. Amelioration of soil acidity and soybean yield after surface lime reapplication to a long-term no-till integrated crop-livestock system under varying grazing intensities. *Soil Res* 2014;**144**:141–9.
274. Pulido Fernández M. *Indicadores de calidad del suelo en áreas de pastoreo*. PhD Thesis. Cáceres: Universidad de Extremadura; 2014.
275. Qadir M, Noble AD, Schubert S, Thomas RJ, Arslan A. Sodicity-induced land degradation and its sustainable management: problems and prospects. *Land Degrad Dev* 2006; **17**:661–76.
276. Rengasamy P. World salinization with emphasis on Australia. *J Exp Bot* 2006;**57**(5): 1017–23.
277. Ma C, Naidu R, Liu F, Lin C, Ming H. Influence of hybrid giant Napier grass on salt and nutrient distributions with depth in a saline soil. *Biodegradation* 2012;**23**:907–16.
278. Brevik EC, Cerda A, Mataix-Solera J, Pereg L, Quinton JN, Six J, Van Oost K. The interdisciplinary nature of soil. *Soils* 2015;**1**:117–29.
279. Yadav S, Irfan M, Ahmad A, Hayat S. Causes of salinity and plant manifestations to salt stress: a review. *J Environ Biol* 2011;**32**(5):667.
280. Metternicht GI, Zinck JA. Remote sensing of soil salinity: potentials and constraints. *Remote Sens Environ* 2003;**85**(1):1–20.
281. Ghassemi F, Jakeman AJ, Nix HA. *Salinisation of land and water resources: human causes, extent, management and case studies*. Wallingford: CAB international; 1995.
282. Mishra A, Sharma SD, Pandey R, Mishra L. Amelioration of a highly alkaline soil by trees in northern India. *Soil Use Manag* 2004;**20**(3):325–32.
283. Bahçeci I. Determination of salt leaching and gypsum requirements with field tests of saline-sodic soils in central Turkey. *Irrig Drain* 2009;**58**:332–45.
284. Vukadinović V, Rengel Z. Dynamics of sodium in saline and sodic soils. *Commun Soil Sci Plant Anal* 2007;**38**(15–16):2077–90.
285. Ayers RS, Westcot DW. *Water quality for agriculture*. Rome: Food and Agriculture Organization of the United Nations; 1985.
286. Cyrus DP, Martin TJ, Reavell PE. Salt-water intrusion from the mzingazi river and its effects on adjacent swamp forest at Richards Bay, Zululand, South Africa. *Water Sa-Pretoria* 1997;**23**:101–8.
287. Howard KW, Mullings E. Hydrochemical analysis of ground-water flow and saline incursion in the clarendon basin, Jamaica. *Gr Water* 1996;**34**(5):801–10.

288. Mao W, Kang S, Wan Y, Sun Y, Li X, Wang Y. Yellow river sediment as a soil amendment for amelioration of saline land in the Yellow river delta. *Land Degrad Dev* 2016;**27**:1595–602.
289. El-Shahway AS, Mahmoud MM, Udeigwe TK. Alterations in soil chemical properties induced by continuous rice cultivation: a study on the arid Nile Delta soils of Egypt. *Land Degrad Dev* 2016;**27**(2):231–8.
290. Zovko M, Romić D, Romić M, Ondrašek G. Soil and water management for sustained agriculture in alluvial plains and flood plains exposed to salinity: a case of Neretva river valley. In: Ahmad P, Azooz MM, Prasad MNV, editors. *Ecophysiology and responses of plants under salt stress*; 2013. p. 473–94.
291. Qureshi RH, Barrett-Lennard EG. *Saline agriculture for irrigated land in Pakistan: a handbook*. Canberra, Australia: Australian Centre for International Agricultural Research; 1998.
292. George S, Harper R, Hobbs R, Tibbett M. A sustainable agricultural landscape for Australia: a review of interlacing carbon sequestration, biodiversity and salinity management in agroforestry systems. *Agric Ecosyst Environ* 2012;**163**:28–36.
293. Szabolcs I. Soils and salinization. In: Pessarakli M, editor. *Handbook of plant and crop stress*. New York: Marcel Dekker; 1994. p. 3–11.
294. Lakhdar A, Rabhi M, Ghnaya T, Montemurro F, Jedidi N, Abdelly C. Effectiveness of compost use in salt-affected soil. *J Hazard Mater* 2009;**171**(1):29–37.
295. Wang L, Seki K, Miyazaki T, Ishihama Y. The causes of soil alkalinization in the Songnen Plain of Northeast China. *Paddy Water Environ* 2009;**7**(3):259–70.
296. Richards LA, United States salinity laboratory staff, diagnosis and improvement of saline and alkali soils. In: Richards L. Agriculture handbook. USA: US Department of Agriculture.
297. Qadir M, Steffens D, Yan F, Schubert S. Sodium removal from a calcareous saline–sodic soil through leaching and plant uptake during phytoremediation. *Land Degrad Dev* 2003;**14**(3):301–7.
298. Yeo AR, Flowers TJ. Salinity resistance in rice (*Oryza sativa* L.) and a pyramiding approach to breeding varieties for saline soils. *Funct Plant Biol* 1986;**13**(1):161–73.
299. Amini S, Ghadiri H, Chen C, Marschner P. Salt-affected soils, reclamation, carbon dynamics, and biochar: a review. *J Soils Sediments* 2016;**16**(3):939–53.
300. Francois LE, Maas EV, Donovan TJ, Youngs VL. Effect of salinity on grain yield and quality, vegetative growth, and germination of semi-dwarf and durum wheat. *Agron J* 1986;**78**(6):1053–8.
301. Akhtar J, Gorham J, Qureshi RH. Combined effect of salinity and hypoxia in wheat (*Triticum aestivum* L.) and wheat-*Thinopyrum amphiploids*. *Plant Soil* 1994;**166**(1):47–54.
302. Saqib M, Akhtar J, Qureshi RH. Pot study on wheat growth in saline and waterlogged compacted soil: i. Grain yield and yield components. *Soil Res* 2004;**77**(2):169–77.
303. Blaskó L. Improvement of saline soils – chemical, biological, land management methods. In: Birkás M, editor. *Soil management and land use*. Budapest: Mezőgazda Kiadó; 2017. p. 308–16.
304. Matosic S, Birkás M, Vukadinovic V, Kisic I, Bogunovic I. Tillage, manure and gypsum use in reclamation of saline-sodic soils. *Agric Conspectus Sci* 2018;**83**(2):131–8.
305. Dang YP, Dalal RC, Buck SR, Harms B, Kelly R, Hochman Z, Schwenke GD, Biggs AJW, Ferguson NJ, Norrish S, Routley R, McDonald M, Hall C, Singh DK, Daniells IG, Farquharson R, Manning W, Speirs S, Grewal HS, Cornish P, Bodapati N, Orange D. Diagnosis, extent, impacts, and management of subsoil constraints in the northern grains cropping region of Australia. *Soil Res* 2010;**48**(2): 105–19.
306. Hanay A, Büyüksönmez F, Kiziloglu FM, Canbolat MY. Reclamation of saline-sodic soils with gypsum and MSW compost. *Compost Sci Util* 2004;**12**(2):175–9.

307. Abdel-Fattah MK, Fouda S, Schmidhalter U. Effects of gypsum particle size on reclaiming saline-sodic soils in Egypt. *Commun Soil Sci Plant Anal* 2015;**46**(9):1112–22.
308. Ahmed K, Qadir G, Jami AR, Nawaz MQ, Rehim A, Jabran K, Hussain M. Gypsum and farm manure application with chiseling improve soil properties and performance of fodder beet under saline-sodic conditions. *Int J Agric Biol* 2015;**17**(6).
309. Rasouli F, Pouya AK, Karimian N. Wheat yield and physico-chemical properties of a sodic soil from semi-arid area of Iran as affected by applied gypsum. *Geoderma* 2013; **193**:246–55.
310. Bennett JML, Cattle SR, Singh B. The efficacy of lime, gypsum and their combination to ameliorate sodicity in irrigated cropping soils in the Lachlan Valley of New South Wales. *Arid Land Res Manag* 2015;**29**(1):17–40.
311. Luo JQ, Wang LL, Li QS, Zhang QK, He BY, Wang Y, Qin LP, Li S-S. Improvement of hard saline–sodic soils using polymeric aluminum ferric sulphate (PAFS). *Soil Res* 2015;**149**:12–20.
312. Nazar R, Iqbal N, Masood A, Syeed S, Khan NA. Understanding the significance of sulfur in improving salinity tolerance in plants. *Environ Exp Bot* 2011;**70**(2):80–7.
313. Modaihsh AS, Al-Mustafa WA, Metwally AI. Effect of elemental sulphur on chemical changes and nutrient availability in calcareous soils. *Plant Soil* 1989;**116**(1):95–101.
314. Slaton NA, Norman RJ, Gilmour JT. Oxidation rates of commercial elemental sulfur products applied to an alkaline silt loam from Arkansas. *Soil Sci Soc Am J* 2001;**65**(1): 239–43.
315. Velarde M, Felker P, Gardiner D. Influence of elemental sulfur, micronutrients, phosphorus, calcium, magnesium and potassium on growth of Prosopis alba on high pH soils in Argentina. *J Arid Environ* 2005;**62**(4):525–39.
316. Yang Z, Haneklaus S, Ram Singh B, Schnug E. Effect of repeated applications of elemental sulfur on microbial population, sulphate concentration, and pH in soils. *Commun Soil Sci Plant Anal* 2007;**39**(1–2):124–40.
317. Amezketa E, Aragüés R, Gazol R. Efficiency of sulphuric acid, mined gypsum, and two gypsum by-products in soil crusting prevention and sodic soil reclamation. *Agron J* 2005;**97**(3):983–9.
318. Singh RK, Redona E, Refuerzo L. Varietal improvement for abiotic stress tolerance in crop plants: special reference to salinity in rice. In: Pareek A, Sopory SK, Bohnert HJ, Govindjee PJA, editors. *Abiotic stress adaptation in plants: physiological, molecular and genomic foundation*. New York: Springer; 2010. p. 387–415.
319. Abrol IP, Yadav JSP, Massoud FI. *Salt-affected soils and their management*. Rome: Food and Agriculture Organization of the United Nations. Soil Bulletin; 1988.
320. Mbau SK, Karanja N, Ayuke F. Short-term influence of compost application on maize yield, soil macrofauna diversity and abundance in nutrient deficient soils of Kakamega County, Kenya. *Plant Soil* 2015;**387**(1–2):379–94.
321. Silva RFD, Bertollo GM, Antoniolli ZI, Corassa GM, Kuss CC. Population fluctuation in soil meso-and macrofauna by the successive application of pig slurry. *Rev Cienc Agron* 2016;**47**(2):221–8.
322. Bohn HL, Strawn DG, O'Connor GA. *Soil chemistry*. USA: John Wiley & Sons; 2015.
323. Choudhary OP, Josan AS, Bajwa MS, Kapur ML. Effect of sustained sodic and saline–sodic irrigation and application of gypsum and farmyard manure on yield and quality of sugarcane under semi-arid conditions. *Field Crop Res* 2004;**87**(2–3):103–16.
324. Gill JS, Sale PWG, Peries RR, Tang C. Changes in soil physical properties and crop root growth in dense sodic subsoil following incorporation of organic amendments. *Field Crop Res* 2009;**114**:137–46.
325. El-Shakweer MHA, El-Sayad EA, Ewees MSA. Soil and plant analysis as a guide for interpretation of the improvement efficiency of organic conditioners added to different soils in Egypt. *Commun Soil Sci Plant Anal* 1998;**29**(11–14):2067–88.

326. Shang Z, Gao Q, Dong M. Impacts of grazing on the alkalinized–salinized meadow steppe ecosystem in the Songnen Plain, China–A simulation study. *Plant Soil* 2003; **249**(2):237–51.
327. Cisneros JM, Cantero JJ, Cantero A. Vegetation, soil hydrophysical properties, and grazing relationships in saline-sodic soils of Central Argentina. *Can J Soil Sci* 1999; **79**(3):399–409.
328. Douaoui AEK, Nicolas H, Walter C. Detecting salinity hazards within a semiarid context by means of combining soil and remote-sensing data. *Geoderma* 2006; **134**(1–2):217–30.
329. Zucca C, Pulido-Fernández M, Fava F, Dessena L, Mulas M. Effects of restoration actions on soil and landscape functions: atriplex nummularia L. plantations in Ouled Dlim (Central Morocco). *Soil Res* 2013;**133**:101–10.

CHAPTER THREE

Soil degradation, restoration and management in abandoned and afforested lands

Teodoro Lasanta[a], José Arnáez[b] and Estela Nadal-Romero[a,*]
[a]Instituto Pirenaico de Ecología, Procesos Geoambientales y Cambio Global, IPE-CSIC, Zaragoza, Spain
[b]Área de Geografía Física, Departamento de Ciencias Humanas, Universidad de La Rioja, Edificio Luis Vives Logroño, Spain
*Corresponding author: E-mail: estelanr@ipe.csic.es

Contents

Abstract

Soil degradation is currently one of the major worldwide environmental problems, being land abandonment one of its main causes. In the last decades an important number of researchers from five continents have studied the interactions between land abandonment and soil degradation. Mediterranean mountain areas and semi-arid environments constitute the areas with the largest amount of abandoned lands, where soil degradation processes act most effectively.

Advances in Chemical Pollution, Environmental Management and Protection, Volume 4
ISSN 2468-9289
https://doi.org/10.1016/bs.apmp.2019.07.002

The hydro-geomorphological evolution of abandoned lands presents variable results depending on the field type (e.g. terraced, sloping, flat), climate conditions (which regulate the revegetation processes), land management and the age of abandonment. In general, if the revegetation processes result in a dense vegetation cover, infiltration increases and runoff and soil erosion decrease. Hence, in sloping fields in mountain areas, the problem is observed in the first years after land abandonment when plant colonization is progressing but not complete. Then, a dense shrub cover protects the soil, reducing both runoff and sediment yield to minimum values. In terraces, the most important erosion process is the collapse of outer walls due to small mass movements, forming scars that facilitate gully development. The lack of maintenance of the drainage systems, created during the cultivation period, favors soil saturation; in these cases, the plant colonization cannot avoid the collapses of the walls. In semi-arid environments, plant colonization is very slow and spatially heterogeneous. Soils tend to crust and are poor in soil organic matter, and consequently the infiltration capacity is very low. Therefore, in some lithologies (as marls and gypsums) soil erosion processes (rills, gullies, piping) are very active, leading in a few years to the total degradation of abandoned fields. All these results suggest that land abandonment should be programmed, and that those already abandoned lands should be managed, emphasizing the role that extensive livestock could produce in these environments.

Keywords: Soil degradation; Land abandonment; Abandoned land; Soil erosion; Afforestation

1. Introduction

Soil degradation is currently one of the most serious environmental problems affecting the World.[1] It consists of a process whereby the ecosystems lose their capacity to provide goods and services to society.[2] Occasionally, the process is irreversible and can lead to desertification.[2,3] The European Parliament identified eight soil degradation processes in the European Union: soil erosion, organic matter decline, contamination, salinization, compaction, soil biodiversity loss, sealing, landslides and flooding.[4] Together with these soil degradation processes are the degradation processes linked to the landscape (homogeneity, fragmentation, reduced biodiversity) and the loss of local culture.[5,6]

There are multiple causes of soil degradation: bio-physical factors combining with socio-economic, technological, demographic, political and cultural ones.[7] In the last few decades, soil degradation has increased due to Global Change: variations in climate, land use and land cover changes, and evolution in socio-economic activities. All the above arises from: (i) more intense agriculture in some areas with higher chemical fertilization and irrigation contaminating the land and water; (ii) land abandonment giving rise to the spread of unmanaged vegetation prone to fires, and at times promoting soil erosion; (iii) expanding urbanization, especially on the coast, which increases land sealing and, therefore, alters the hydrological cycle; (iv) deforestation in some areas causing loss of diversity and fragmenting habitats; and (v) changes in rainfall patterns and more frequent and intense droughts affecting vegetation and soil erosion.[8–10]

Land abandonment is one of the main causes of soil degradation[11–14] and loss of ecosystem services, jeopardizing the sustainability of some societies.[6,15] Ramankutty and Foley[16] calculated that approximately 1.5 million km^2 have been abandoned worldwide, a figure that has risen considerably in the 21st century, and is forecasted to continue for the next few decades.[17] For Strijker,[18] it is a socio-economic problem involving the environment at global, regional and local scales. Land abandonment can be defined in various ways depending on the affected territory, the context of the study and its environmental or socio-economic implications.[19] Baudry[20] defines land abandonment as "the complete termination of land use and management, including a change toward less intensive models in land use, or as the complete termination of its use, abandoning it to spontaneous dynamics". According to Pinto Correia[21] "land abandonment is applied to land where traditional or recent farming is no longer carried out". Martínez Hernández,[22] taken from Romero Díaz,[23] defines land abandonment as "the phenomenon of permanent cessation of agricultural activity in a specific area". Alonso-Sarriá et al.[24] defined agricultural abandonment as "the consistent and indefinitely cessation of farming activity in a particular area, and without any recent attempt to resume this activity or any other profitable activity". Most of the literature studies the land abandonment, which frequently leads to an expansion of vegetation due to plant succession and a land use change, usually for livestock and/or forestry. For Baudry[20] and Pinto-Correia[21] the concepts of "land marginalization", "land abandonment" and "land extensification" are synonymous. However, for other authors, the abandonment of agricultural land is a complex and gradual process, which starts with progressive marginalization (no management), first leading to extensification, and later to abandonment, a stage where a change in plant cover takes place due to secondary succession.[17,25] In this chapter, abandoned lands are understood as those that have undergone this evolution through the marginalization, extensification and abandonment stages.

Soil degradation in abandoned lands is a complex issue, since several factors interact depending on the climatic and edaphic conditions, the characteristics of the area where the abandonment occurred, the topography, the management before and after land abandonment, and especially in the case of agricultural land, the field types (terraced, sloping flat). In addition, it is a subject requiring a certain time perspective, since a large part of soil degradation/conservation depends on how long ago abandonment took place and the capacity of the soil to facilitate the stages of succession in plant recolonization.[26]

The following pages consist of a review of soil degradation relating to land abandonment, focusing particularly on the main areas of abandoned lands in Europe: Mediterranean mountain areas and semi-arid environments. First, there is information on the literature on abandoned lands and an explanation of the spatial and temporal relationships. The last section emphasizes the need to manage abandoned lands to control its effects on soil degradation.

2. Land abandonment in the scientific literature

The large surface area of abandoned lands and its implications for the environment, landscape and for society and economy, led to their being studied. On March 1st, 2019 a bibliography research was carried out in the Scopus database. The words "land abandonment" and "abandoned agricultural" were added. 1067 and 592 documents were found, respectively. Crossing both keywords showed that 21 documents were repeated, so a total of 1638 were used in the bibliometric analysis.

The first paper in the Scopus database was published in 1928,[27] and made particular reference to issues relating to medicine. The second was published in 1934[28] and deals very marginally with soil erosion processes in abandoned fields. The third[29] shows the flora composition in the first years after abandonment. The next publication was not until 1978,[30] analyzing the nitrogen export in 21 catchments in Canada over 25 months. A good correlation was found between exported nitrates and the expansion of abandoned crop fields. From 1978, at least one document per year has been published (Fig. 1). In the 1980s, less than 10 documents a year were published, which rose to 10–20 a year in the 1990s. From 2008, a minimum of 70 documents a year have been published. After 2010, this rose to over 100 a year, and recently, at least 1 document is published every 3 days, suggesting the significant impact of the subject in current science (Fig. 1).

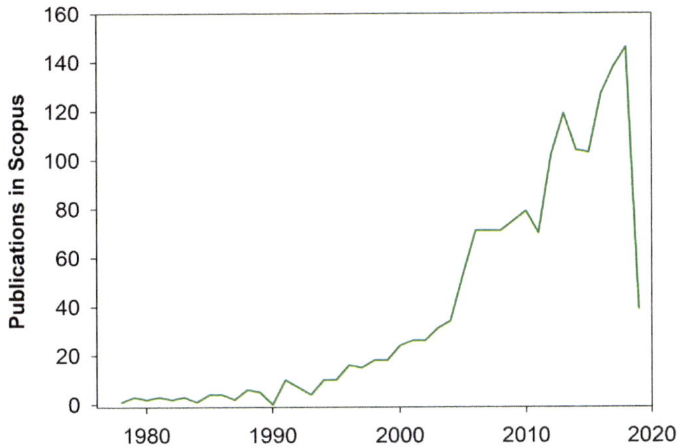

Fig. 1 Number of publications in Scopus database with the term "land abandonment" and "abandoned agricultural". Search date: March 1st, 2019.

Given the huge diversity of subjects relating to land abandonment, the documents use more than 350 keywords. If keywords are taken to be used at least 25 times, the number of keywords rises to 90, of which 15 refer to locations in the study area: Europe (195), Spain (163), Eurasia (156), United States (71), Italy (67), Southern Europe (61), Mediterranean Region (57), Mediterranean (57), Mountain Region (53). Fig. 2 includes information on the most used keywords. Abandoned land, land abandonment,

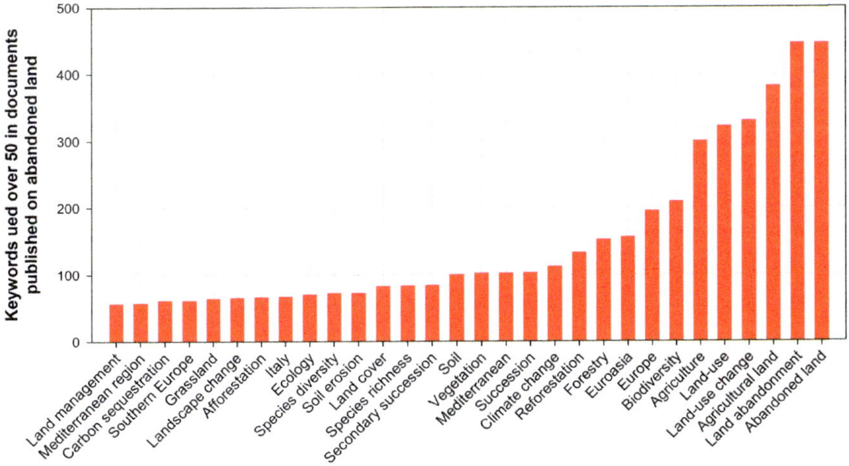

Fig. 2 Keywords used over 50 times in documents published on abandoned land in the Scopus database. Search date: March 1st, 2019.

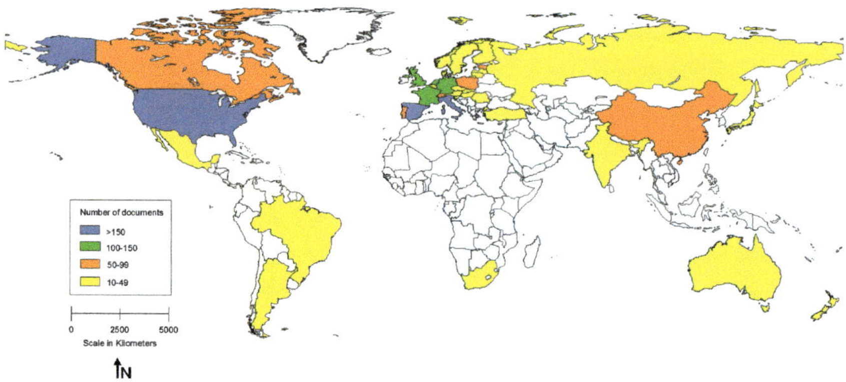

Fig. 3 Worldwide distribution of studies related to "land abandonment" and "abandoned agricultural" terms.

land use change, land use, agricultural land, agriculture have been used more than 300 times. The following are repeated at least 100 times: reforestation (133), climate change (112), biodiversity (209), succession (103), vegetation (102), soils (100). Other subjects often studied in the literature are secondary succession (84), land cover (82), ecology (70), among others.

Scopus also includes information on country/region that brings us closer to where the work was carried out (see Fig. 3). The documents come from the 5 continents, with the highest number from Europe (1629), followed by North America (418), Asia (191) and Latin America (132). 56 deal with Africa, and 54 come from Oceania. The study area is not known or not specified in 39 documents, and most of them are global reviews. Table 1 shows

Table 1 Countries with more than 10 documents. Source: Scopus. Search date: March 1st, 2019.

Country:	No	Country:	No	Country:	No	Country:	No
United States	341	Spain	309	Italy	160	Germany	145
Netherlands	123	France	118	United Kingdom	105	Switzerland	88
Canada	77	China	69	Portugal	57	Poland	50
Estonia	50	Belgium	46	Japan	41	Brazil	40
Australian	38	Russian Federation	38	Czech Republic	36	Sweden	35
Austria	33	Denmark	27	South Africa	27	Mexico	27
Hungary	23	Finland	23	India	20	Latvia	19
Romania	17	Argentina	16	Slovakia	15	Slovenia	15
Lithuania	15	Norway	12	Turkey	11	New Zealand	11

the most studied countries. It can be seen that the more developed countries have the most documents, coinciding with the areas with the most abandoned land. United States (341 documents) has the most, followed by Spain (309), Italy (160) and Germany (145). However, recently there has been a considerable increase in the number of publications in some other countries. These are certain eastern European countries (Poland, Estonia, Czech Republic ...), where post-socialism has led to a great deal of land being abandoned[31] and, on the other hand, China (69 documents) and India (20).

A point we would like to highlight is the large number of documents published on the Mediterranean region, shown by the many times that various terms (Mediterranean region, Mediterranean environment, southern Europe) and names of Mediterranean countries (Spain, Italy, France, Portugal, Romania, etc.) appear in the keywords and study areas (see Figs. 2 and 3, and Table 1). In this region, land abandonment has been happening for many years and over an extensive area, both in very steep slopes cultivated in the past and in the semi-arid land prevailing in several regions, giving rise to problems of fragile soils with low fertility, which are aggravated by climate change.[24]

3. The spatial and temporal process of land abandonment

There is relatively little literature on the spatial and temporal process of land abandonment on a wide scale, perhaps due to the difficulty in obtaining accurate information. The official statistics on the change in the number and size of farms can give an approximation of the amount of abandoned land, but not the location of the abandoned fields. Statistical information on land uses and land covers has the same limitations. The temporal process of abandonment and its spatial distribution can be found from graphic documents: maps of land uses and land covers, photographs, aerial photographs, orthoimagery or satellite images. However, it must be remembered that plant succession takes place after abandonment, which removes traces of the cultivated phase more or less rapidly, depending on the environmental and management conditions, as shown in a later section. Thus, reliable information can only be obtained from documents close to the age of abandonment or, at least, before the plant cover removed the field edges. This fact means that using satellite images, which can provide rapid information over large areas, is restricted to land abandonment recorded during the last few decades,[32,33] the same as with orthoimages giving highly detailed information, but requiring a great deal of mapping and field work to complete a supramunicipal area. A temporal sequence of aerial photographs constitutes the best document to find the age of abandonment and its spatial distribution, but demands a great effort in photo-interpretation, and therefore, there are hardly any study covering an area larger than a district or mountain valley.[34]

The above limitations make the literature focus on recent abandonments, no more than fifty years, which coincides in several countries with industrialization, urbanization and intensification in the lowlands,[35,36] which favored depopulation of regions. However, in many places - especially in mountain areas - land was abandoned much earlier.[37,38] This can be synthesized in three different stages, with different characteristics and spatial boundaries: early land abandonment in the mountains, the removal of land linked to the Common Agricultural Policy (CAP), and recent abandonment in Eastern bloc countries. The first and third were spontaneous abandonments, connected to the sudden or gradual collapse of the societies supporting active land management, whereas the second is induced by policies aimed at regulating markets for certain agricultural products.[38]

3.1 Land abandonment in mountain areas

The first abandonments in mountain areas recorded in the literature in Europe occurred in the early 19th century in the French Alps,[39] in Hérault (southern France)[40] and the southern Alps.[41] In the mid-19th century, abandonment affected some of the regions that were becoming industrialized in the United Kingdom, Germany, Switzerland, north Italy and France, where many farmers emigrated to the cities, thus reducing the pressure on farmland.[42] In the east of North America, farming decreased in the 1880s and the area of agricultural land fell from 187,900 km^2 to 48,800 km^2 between 1880 and 1997, spiking sharply in the 1940s.[43] The process spread to several other countries from the first few decades of the 20th century, and intensified in the more developed ones in the mid-20th century, when large tracts of agricultural land were marginalized and abandoned.[42]

There was a general farmland abandonment in all the mountain areas of western Europe in the middle of the 20th century, although it was unevenly distributed. The Mediterranean mountains recorded more abandonments than Alpine areas, which had more resources to adapt to the new management systems.[34,37] Wessteiner et al.[33] and Alonso-Sarriá et al.[24] included large amounts of the literature on land abandonment in the Mediterranean basin in their research. However, there are not many studies quantifying the surface area of abandoned land: Mallet[44] in the Briançonnais Alps and Réparaz[45] in the Provence Alps estimated that 79% and 97% (respectively) of total farmland was no longer cultivated. Darbellay[46] noted the rapid shrinkage of agricultural land in the mountains of Switzerland, where the number of farms fell from 68,000 in 1955 to 46,000 in 1980. Haefner and Günter[47] calculated that abandonment ranged between 44% and 64%, depending on the Swiss canton. In the Oô valley (French Central Pyrenees), 84% of farmland was abandoned from the start of the 20th century up to the beginning of the 1980s.[48] The extent of abandonment has also been observed in the mountains in Poland, Slovakia, and the French Prealps, where Kozak et al.,[49] Kuemmerle et al.[50] and Taillefumier and Piégay,[39] respectively, estimated that 20% of farmland had been abandoned. This rose to 70% in the Eastern Alps,[51] while it was 30% in the Carpathians.[52] There are several studies quantifying agricultural abandonment in mountain areas of the Iberian Peninsula, a process that may have begun at the end of the 19th century and reached its peak from the 1950s to the 1980s.[53] Lasanta[54] notes that in the Central Spanish Pyrenees 71% of farmland has been abandoned, 99% in Cameros Viejo (north-western Iberian Range, Spain),[55] and 70% in the Alpujarras (Penibetic Range, Spain),[56] to give just a few examples.

3.2 Land retirement in the European Union due to the CAP (1988–2008)

Farmland abandonment in the flat areas of the developed countries was not intense for most of the 20th century, because there were more resources allowing the agricultural sector to adapt to the new demands of the market.[34] In addition, the lowlands benefited from better transport and infrastructure, as well as economies of scale. However, from 1988, the EU, through the CAP, encouraged marginalization and abandonment of land by supporting extensification of production (Regulation EEC no 4115/88, modified by no 708/92 and no 838/93), retirement of arable land (Community Rule no 1034/88 and EEC Regulation no 1765/92), and early retirement (EEC Regulation no 2079/92), among other rules and regulations established up to 2008. Therefore, this abandonment was at the hands of agents outside the rural environment.[38]

CAP reform in 1992 established permanent land retirement of farmland in areas with a strong risk of environmental degradation, or which were to be preserved by special management, or temporarily (set-aside) (5 years maximum), in order to control production to prevent surpluses. Farmers received a subsidy to set-aside part of their land, which was voluntary for those producing less than 92 tonnes of cereal, and compulsory for those producing more, who had to set-aside at least 20% of their land. The percentage decreased slightly if different fields were not cultivated each year.

The EU land retirement programs had precedents within and outside Europe. In the United States, the agricultural policy, established in the Agricultural Adjustment Act (1933), encouraged programs aimed at preventing food surpluses to stabilize prices and maintain farmers' incomes. Depending on the market conditions, the retired areas varied from year to year - in 1992 it was 7.5 million ha, but 25 million in 1987.[57] In Canada, the Lower Inventory for Tomorrow (1970) aimed to remove 8.9 million ha from production, and offered farmers financial compensation to do so. In 1969, the Finnish Land Reserve Program was introduced to stop production on 350,000 ha.[58]

There is little information for the EU as a whole on land retirement due to the CAP. Feranec et al.[59] used data from the Corine Land Cover to analyze changes in agricultural use (intensification and extensification) in 24 countries in Europe between 1990 and 2000, noting that the change in use affected 88,000 km^2, equivalent to 2.5% of the total surface. The greatest changes took place in Portugal (9.9% of total farmland), Ireland (7.9%) and the Czech Republic (6.4%), while Germany, Austria, Slovenia, Bulgaria and Poland recorded less than 1% change. IEEP and Alterra[60] used the same source of information to try to locate the land abandoned between 1990 and 2000, reaching the conclusion that land use changes recorded at that time did not coincide with what the literature considered to be "hot spots" of land abandonment based on regional studies. Thus, the Pyrenees, Central Massif in France, the Austrian Alps, mountain areas in Germany and most of the 24 mountain areas that Macdonald et al.[61] indicated as "hot spots" of land abandonment remain outside. IEEP and Alterra[60] attributed the main cause of the discrepancies between the data in Corine Land Cover and case studies to the difficulties in interpreting land cover from satellite images. Verburg and Overmars[61] indicated the complications involved in distinguishing abandoned agricultural land from pastures by remote sensing. Pointereau et al.[17] stated that land abandonment should be studied on a regional scale from case studies, as on a country-wide scale some abandoned areas could be masked by new plowing in others.

Pointereau et al.[17] studied the change in Utilized Agricultural Area (UAA) between 1990 and 2000 in France, Spain and Portugal and they concluded that overall, 3.3 million ha had been abandoned, a loss of 2% of the UAA in France, 4% in Poland and 8% in Spain. Keenleyside and

Tucke[25] noted that, in four regions in Finland between 1990 and 2000, extensively managed areas recorded more abandonment and lying fallow in the long-term, reaching up to 39% in Pohjanmaa, a region engaged in extensive livestock farming, while in arable land managed intensively, abandonment ranged between 4% and 16%.

3.3 Land abandonment in the Soviet sphere

In Central and Eastern Europe, the political and socio-economic actions caused by the Communist regime led to changes in land use and the landscape throughout the 20th century and beginning of the 21st.[31,62,63] During the time of the Soviet period (1945–90) large collective farms were created in the most fertile areas, while marginalized ones were used for forests for recolonization, which simplified and homogenized the landscape.[64] After 1989, agrarian reform was carried out by returning the land to its former owners, who frequently lived in cities. However, the lack of interest in agriculture or of the capital required to form an agribusiness led to many farms discontinuing.[65,66] Abandonment was also boosted by the collapse of collective farms, retirement of many farmers, and the young people emigrating from the country to the cities or to other countries.[67]

Large areas of land were abandoned within just a few years. Van Dijk et al.[68] stated that between 10% and 20% of agricultural has been abandoned in Central and Eastern Europe. Kuemmerle et al.[50] studied land abandonment in an 18,000 km^2 triangle in the Carpathians on the borders with Poland, Slovakia and Ukraine. They used satellite images to quantify the fact that, between 1986 and 2000, 20.7% of agricultural land in Slovakia, 13.9% in Poland and 13.3% in Ukraine were no longer cultivated. They observed that, in Poland, a larger proportion of land that was previously collectivized had been abandoned, double the amount that remained in private hands. This was in answer to the lower profits from agriculture and changes in ownership during the communist years and the agrarian reforms after 1989.

Muller and Kuemmerle[69] calculated that 21% of land in the District of Arges (Romania) was abandoned between 1990 and 2005. Other studies showed the high proportion of farmland abandoned after the Eastern bloc fell apart. Thus, Keenleyside and Tucker[25] stated that, in Poland, 17.6% of agricultural land was abandoned, with a sharp increase between 1998 and 2002, and at rates of 100% in the provinces of Mazovia, Lubelskie, Podkarpackie and Podlaskie. In Estonia, 10.1% of farmland was abandoned.[70] In Latvia between 1989 and 1999, 10.3% of agricultural land was abandoned;

the less fertile areas were no longer cultivated, small plots and those affected by bad weather, especially in regions like Latgale (21.1% of crop land abandoned), and other smaller ones where abandonment rose to as much as 50%.[62] Ruskule et al.[71] studied the process of abandonment in the highlands of Vidzeme (Latvia), and calculated that, from 1989, between 25% and 35% of farmland was no longer cultivated, depending on the municipality. Kozak et al.[49] observed that a third of the cultivated area in the Beskid Maly mountains (Carpathians, southern Poland) were abandoned between 1965 and 1997, with the process not yet finished. Hostert[52] gave rates of abandonment of up to 56% in some areas of the Carpathians, and around 30% in the whole of the mountain range after the fall of the Iron Curtain. In some regions of Latvia, up to 50% of arable land was abandoned between 1989 and 1999.

3.4 Land abandonment of cultivated lands in Europe in the coming decades

Given the complex reasons intervening in land abandonment, it is difficult to know whether the coming decades will bring further abandonment, and how to quantify and locate it. However, several studies have been made based on complex computerized models (e.g. Common Agricultural Policy Regional Impact Assessment, CAPRI, and Dynamics Conversion of Land Use and its Effects, Dyna CLUE) aiming to forecast land use changes in Europe, and which lands will be abandoned in the near future.[25,61,72,73] The scenarios take particular account of the level of globalization and regulation of economic activity.

The results from all the models indicate that there might not be very much land abandoned in the next 20–30 years. However, there are large variations among the models, and within the same model, depending on the scenarios included. The most optimistic ones indicate that 0.7% of arable land in Europe will stop being cultivated,[72] while others raise that figure to 6.7% in 2030.[61,74] Most abandonments will occur if there is strong global competition among agricultural products and little aid from the CAP for extensive farming. Nevertheless, some scenarios indicate that abandonment will still take place even though global competition is reduced, crops are heavily regulated, and there is abundant support for agriculture and the environment. Estimated abandonment is reduced in scenarios including high levels of biofuel production in the EU. However, some authors think that this is a very optimistic perspective, as biofuels are expected to be produced outside Europe, despite the incentive from the EU Directive on Renewable Energy.[25]

The models agree fairly consistently with the fact that areas at greatest risk of abandonment are located in Finland, Sweden, the Pyrenees, north-west Spain, north Portugal, the Massif Central in France, the Apennines, the Alps, the German high land near the border with the Czech Republic, and the Carpathians. Most of the land at severe risk are in mountains at high altitudes and coincide with those classified as High Natural Value (HNV). Paracchini et al.[75] and IEEP and Alterra,[60] indicated that 19.8% of arable land, and 28.1% of pastureland included in HNV areas could be abandoned. Non-irrigated land in arid and semi-arid zones may also be abandoned. In this respect, Nainggolan et al.[76] predicted that some non-irrigated fields will remain abandoned in the south-west of Spain, especially if there are subsidies for reforesting agricultural land.

Estimates of future land abandonment must be viewed with a great deal of caution, since they are limited by the information available, as well as by uncertainty with policy decisions and future socio-economic development. In addition, the models are fairly deterministic and assume that land owners will respond immediately to external incentives, especially financial. However, reality shows that farmers frequently take into account other social and cultural issues, leading them to farm the land even though there is little profit on it; sometimes they continue at a financial loss to complement income obtained from other activities. Keenleyside and Tucker[25] stated that both scenarios predicting large-scale abandonment - as they assume high levels of market liberalization and weak environmental regulation, which are probably unrealistic premises - and those inclined to underestimating future abandonment - as they do not take into account the hypothetical recovery of regulations on compulsory land retirement (in force from 1988 to 2008), the effects of soil erosion, rural depopulation and climate change - must be rejected. The authors of this paper estimated that, in a realistic scenario, about 3—4% of arable land will be abandoned, equivalent to 126,000—168,000 km^2 by 2030, with marked variations among regions, although HNV areas are under most threat.

4. The impact of land abandonment and reforestation on soil degradation and hydrological response

4.1 Plant succession in abandoned lands

Vegetation cover is possibly the factor that best explains soil degradation in abandoned land, more than rainfall intensity or the slope.[26,77]

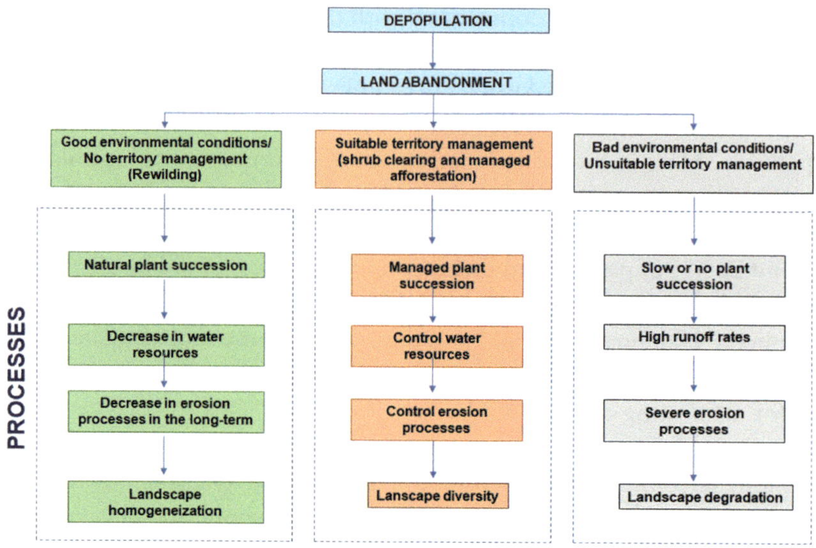

Fig. 4 The impact of land abandonment and land management on soil degradation: revegetation processes, runoff and soil erosion response.

Vegetation directly impedes erosion by protecting the soil surface from rainfall splash, and indirectly by improving the quality of the soil (fertility, porosity, structure of the aggregates …), which helps water to infiltrate and reduces runoff.[78] Therefore, this section explains the plant succession process in abandoned lands (Fig. 4).

The first effect of land abandonment is the start of secondary succession or plant colonization,[79] which has important implications for the soil (quality, conservation, carbon sequestration) and hydrology at different spatial scales, not forgetting changes in landscape structure and aesthetics, biodiversity, the fire regime and available pasture.[34]

Secondary succession is a highly complex process, as it depends on multiple natural or biophysical factors and those of human origin or social factors.[80,81] Among the first ones are edaphic characteristics (soil depth and fertility), hillslope topography (exposure, differences in height, altitude and shape: concave, convex or straight), climate (temperature, rainfall, evapotranspiration), the lithology determining the flora composition, and distance to natural plant cover. Among the anthropogenic factors are the age of abandonment, management when cultivated and after farming ceased, especially livestock husbandry and the use of fire.[82]

After a crop field has been abandoned, rapid colonization by herbaceous plants takes place, usually weeds that were already present in the soil, and ruderal on the edges of the field and nearby paths. Within very few years (3–5), depending on the soil fertility and climate conditions, the field can be colonized by a very dense herbaceous cover forming pasture, at least in humid or sub-humid environments. In terraced slopes, it has been observed that plant colonization is very rapid if the slope or face was covered with vegetation when cultivated, as it is facilitated by the proximity of the seed bank and greater depth of soil than in sloping fields (Fig. 5).[83] On the other hand, in Mediterranean and semi-arid environments, the process can be delayed, with colonization by herbaceous plants being very slow and unevenly distributed.

The use of extensive livestock farming is one of the aspects conditioning the rate of plant succession. With a suitable number of livestock, abandoned fields act as pasture for decades and delay colonization by shrubs and trees. If there is overgrazing, plant succession is interrupted, gaps appear and the soil compacts, leading to reduced infiltration and increased runoff and soil loss through erosion processes.[84] Mitchley and Ispikoudis[86] and Zaravali et al.[86] suggested that the expansion of shrubland in Europe after the Second World War can be explained by the decrease in goat grazing, by not burning shrubs to facilitate the regeneration of pasture, and by no longer using them for fuel and charcoal. If livestock does not interfere with plant succession, the

Fig. 5 Abandoned terraces (Iberian System, Spain). This hillslope with calcareous substrate was cultivated with cereal until 1970s. Nowadays, a dense vegetation cover predominates with *Genista scorpius*, *Buxus sempervirens* and *Juniperus communis*. Photo. T. Lasanta, April 2019.

shrubland rapidly spreads and covers fields within a few years.[51,61,87,88] The above is a summary of the huge complexity of interactions between grazing and plant succession, as indicated by Chauchard et al.[41]

The persistence of a dense shrub cover depends on the colonized species, the frequency of fire and soil quality. In the Spanish Pyrenees, it has been observed that clusters of gorse die after 50–60 years, opening gaps in shrub stratum that allow herbaceous plants to enter (*Brachypodium pinnatum*, *Carex flacca*, *Bromus erectus*, *Dactylis glomerata* ...) or other scrub and shrubs (*Buxus sempervirens*, *Juniperus communis*, *Crataegus monogyna*, *Rosa* sp.).[89] Fire is an element used traditionally to remove shrubland and promote the regeneration of grazing land, at the cost of interrupting plant succession.[90,91] Soil quality is another determining factor in the rate of plant succession, with it being very rapid in deep soils rich in organic matter. However, a cover of shrubs can remain unchanged for decades in abandoned fields with thin, eroded or poorly structured soil, even though there is sufficient rainfall for trees to grow and they are close to forest (see Fig. 6).[61,92,93] Under these conditions, progress from shrub cover to one of trees is very slow or never reached.[94,95]

In arid and semi-arid environments, precipitation (volume, intensity and seasonal distribution) and infiltration rates determining soil moisture are highly important factors. Lesschen et al.[96] found that recovery of vegetation in abandoned fields in the south-east of Spain was slow, especially on loamy substrates (Fig. 7), due to low infiltration rates. Usually, when there is little

Fig. 6 Severe sheet wash erosion. Until the middle of the 20th century, this hillslope located in the Central Spanish Pyrenees was cultivated with cereals. South-facing slopes involve a high water stress that hinders plant colonization, especially in the steepest fields where sheet wash erosion occurs. Photo. T. Lasanta, 1995.

Fig. 7 Abandoned lands in the Southeast of Spain. In marly lithologies, plant colonization is a slow process, emerging crusts in the bare soil. In the foreground, two small depressions due to surface collapses associated to piping processes. Photo. E. Nadal-Romero, September 2009.

rainfall, plant colonization takes place on patches of vegetation alternating with baer soil.[12,97] Ries et al.[98] found that on gypsum soils in the Ebro Depression (Spain), plant succession is very slow, so that 6–8 years after abandonment it has not reached 50%, and there are even fields abandoned a long time ago that hardly reach 60%. These authors attributed difficulties in the spread of vegetation to low water content and poor soils. Haase et al.[99] indicated that land abandonment in arid areas causes colonization by a single species of plant which remains unchanged for decades; in the south-east of Spain, *Anthyllis cytisoides* L. and *Retama sphaerocarpa* Boiss form rows or tiers that alternate with baer patches. Pugnaire et al.[79] concluded that plant colonization on abandoned land in semi-arid environments depends heavily on the availability of seeds and climate variability. These authors observed that cultivating the land for centuries may have restricted the availability of seeds from some perennial species, and that germination is better in relatively wet years.

The change from shrub to young forest can be rapid (20–30 years) in wet areas of Atlantic and continental Europe,[40,100] or slow (more than 80 years), as in the Mediterranean mountains.[101,102] An important factor in reaching the forest stage is the proximity of abandoned fields to forest areas (Fig. 8).[101] Verdú and García-Fayos[103] indicated the importance for revegetation of trees being near to the abandoned fields. They have an indirect

Fig. 8 Abandoned terraces in the Central Spanish Pyrenees. They were cultivated with cereals until the middle of the 20th century. The presence of *Quercus gr. Faginea* at the edges of the fields has favoured a fast revegetation process. Photo. T. Lasanta, November 2018.

influence by acting as "perches" where birds can rest and let seeds fall to the ground, either by regurgitation or defecation; and directly, by improving micro-environmental conditions underground by maintaining in a better way soil moisture and increasing porosity, which helps seeds to germinate.

Plant succession in abandoned fields is a dynamic process in which the natural vegetation starts to invade the space from where it was removed. This evolution is scaled into a series of stages ranging from pioneer communities to other, more structured and complex ones. Clements's models suggest that succession in abandoned fields follows a linear progression, starting with herbaceous plants that then give way to shrub and, later, trees.[104,105] However, as we mentioned before, many biotic and anthropogenic factors interact in the process, giving rise to a wide variety of revegetation patterns[106,107] affecting soil degradation and hydrological dynamics.[38] This is seen especially in the Mediterranean region of Europe, due to the large area of abandoned land, the presence of different field types, and the complexity of plant succession models for scenarios of scant and irregular precipitation.[12,23,79,82,87,92,97,101–103,107–109]

4.2 Changes in soil quality

Soil quality is the capacity of the soil to sustain a specific natural or agricultural ecosystem over time.[110] When a field is abandoned, the soil is often poor quality, low fertility and very stony. Casalmiglia et al.[111] found that soils on recently abandoned terraces had little organic content and a low microbial mass, a legacy from its agricultural days. On abandoned land, soil quality alters depending on the relationship between vegetation and erosion. When abandonment is followed by revegetation, which may change the land to forest, it generally helps to improve soil quality.[39,40,51,112] When, on the other hand, there is no plant succession, continuous degradation occurs with heavy soil loss through erosion and, therefore, a loss in soil quality.[113] Land abandonment can result in recovery of the previous geo-ecosystem as vegetation and soil quality improve.[114] However, no clear patterns in soil properties have been observed in the literature. In general, significant changes in the physical and chemical properties, and biochemical soil cycles are reported.[115–117] Improvement in soil quality means a higher content of organic matter, less apparent density and greater aggregate stability, leading to changes in soil hydrology: increased infiltration, improved conductivity and a better capacity to retain water.[118] On the other hand, no vegetation cover after abandonment can cause the opposite effects, with a gradual loss of nutrients and organic matter, and crust formation increasing runoff.[119]

Vegetation recovery following abandonment is relatively rapid in fertile soils with a humid or sub-humid climate.[120,121] However, it is very slow in arid and semi-arid areas, where water is a key resource in the development of plant cover.[122] Spatial redistribution of water is an essential factor in understanding plant cover patterns and soil degradation.[12] Where soils have a better infiltration capacity, plant cover develops, with a positive feedback between water and vegetation.[123] Higher content of organic matter is usually found under the plants, which facilitates soil aggregation and increases porosity by raising infiltration rates.[12,124] On the other hand, the infiltration capacity in areas of bare soil is often very low, and linked to heavily compacted soils and surface crusts.[125] Surface crust is frequently formed on soil in arid and semi-arid environments.[13] The combination of large quantities of calcium carbonate in the soil and high rates of evaporation after rain leads

Fig. 9 Biological crust in gypsums in the Central Ebro Depression (Spain). Photo. C.L. Alados.

to the formation of crusts in areas with no vegetation and not plowed.[38] According to Pueyo,[126] there are two types of crust: the physical, created by the impact of rain drops on a bare surface, and the biological formed by lichens and algae (Fig. 9). The physical crust obstructs pores in the soil, making infiltration difficult and promoting surface runoff.[127] The hydrological function of the biological crust is less clear, and has contradictory results on its role in infiltration (see 126). The physical crust is also caused by livestock trampling and compacting the soil, reducing infiltration and increasing runoff and soil erosion.[85,128]

Degraded soil displays negative feedback, where the absence of vegetation leads to very low infiltration rates that limit plant colonization. Linked to this feedback are others, such as poor local capacity to absorb water, which increases runoff over wider areas and soil degradation through erosion.[12,129] Soil degradation frequently becomes irreversible and the only way back is by taking measures to restore the ecology and hydrology (Fig. 10).[130]

Abandonment of agricultural land generally leads to an increase in organic carbon in the soil, resulting in soil quality improvement.[131] However, although agricultural land abandonment is a widespread phenomenon, detailed studies on carbon dynamics after abandonment are less noticeable and few data are available, and opposite results can be found in the literature (see Fig. 2). In general, literature indicated that woody encroachment on former croplands in mid mountain areas produce an increase of the original

Fig. 10 Hydrological restauration in Bardenas (Spain). To prevent the progress of a small gully, a small wall was constructed and the incipient gully was filled with soil. Note that in the lower part of the wall, a small outlet to runoff has been created. Photo. T. Lasanta, November 2016.

soil organic carbon.[132] Gabarrón-Galeote et al.[133] using a chronosequence approach of land abandonment, indicated that in the wettest sites (>650 mm) soil organic carbon increased after land abandonment with a mean accumulation rate about 0.11 kg C m^{-2} y^{-1}. However, in the driest sites no changes were observed. Novara et al.[134] showed that abandonment increases soil organic carbon (SOC) stock by 9.03 Mg C ha^{-1} on average. However, other studies concluded that plant succession after land abandonment does not produce an increase of SOC sequestration.[135] Recently, based on these contradictory results, Nadal-Romero et al.[136] suggested that further studies are required to understand the effects of land abandonment on soil properties and SOC dynamics, mainly due to the relevant role played by soils in the sequestration of soil organic carbon (see 137).

4.3 Implications of land abandonment in soil erosion and sediment transport

As already seen, land abandonment occurs under very diverse situations in respect of environmental conditions, field type and age of abandonment, and thus the hydromorphological response is very varied. In addition, it must be remembered that the soil in abandoned fields is normally shallow and only slightly developed.[138] In some cases, this is due to their location on very steep slopes or in areas with little rainfall; in others, because past farming methods did not conserve the soil, as shifting agriculture practices.[139] Since the soil is shallow, the organic matter and most nutrients lie near the surface, where the soil is especially vulnerable to erosion. The loss of a small surface layer can be very detrimental to soil degradation.

In most cases, a gradual increase in the density of the plant cover is observed after abandonment, generally meaning that soil erosion has decreased from when it was cultivated.[38,140] This is a logical consequence of the protective effect provided by vegetation against splash, the loosening of particles and impeding runoff evolution. Moreover, as mentioned before, soil properties tend to improve after some years of abandonment and facilitate water infiltration.[141,142] In abandoned fields in the south-east of Spain, it has been observed that, in a short space of time (4—10 years), there is a significant increase in the content of organic matter, aggregate stability, permeability and water retention capacity.[143] In the Greek island of Lesbos, Kosmas et al.[142] studied fields on hillsides abandoned for 40—45 years with slopes between 18 and 25%. They found that, in soil developed on lava, the content of organic matter increased from the 0.9% when cultivated, to 2.4% following abandonment. In the Maestrazgo (Castellón, Spain), it was found that land abandonment improved the content of organic matter in the soil and aggregate stability, with a significant increase in their size.[144] Nadal et al.[145] also observed an increase in soil fertility as the plant cover advanced. These authors pointed out that it is not a linear process, but very much affected by post-management after abandonment, especially disturbances such as fire, topography and fertility before abandonment. In fields with good fertility when cultivation ended, an initial decrease in the levels of fertility could be found, needing over 50 years to recover. Koulouri and Giourga[108] observed that the density of aggregates increased after abandonment, while porosity decreased from when the land was cultivated.

Soil erosion decrease with the increasing age of the abandonment, as the plant cover advanced and became denser, with a more complex structure. This seems to suggest that the problem lies in the early years of abandonment, when plant colonization advances gradually, but the cover is not complete.[146] For example, Cerdà et al.[147] suggested that in the short-term (2-years) an increase in soil and water losses was observed in abandoned fields; however, in the long-term (11-years) land abandonment resulted in lower erosion rates. However, the subject is much more complex, since sometimes fields that have been abandoned for longer are those with the most degraded soils and severe erosion. In the Aísa valley (Spanish Pyrenees), Ruiz Flaño et al.[148] found that fields abandoned in the first few decades of the 20th century were highly degraded, with a prevalence of micro-environments paved with stones resulting from the fine particles having migrated from the soil, incisions and small mass movements. On the other hand, fields abandoned in the 1950s and 1960s were better preserved, with weak, widespread stream formation, both on herbaceous and shrub surfaces. Poesen et al.[149] stated that the older abandoned fields have a greater percentage of stones caused by selective erosion from runoff dragging away fine particles and leaving stones due to their higher weight.

This poorer condition of older abandoned fields can be interpreted by the following argument: the first fields to be abandoned were the worst (steeper, poor soil, low fertility and very stony), cultivated at a time of heavy demographic pressure and with few or no methods of conservation (frequently using shifting agriculture practices), so plant colonization after abandonment was difficult (Fig. 11). These conditions gave rise to increasingly active erosion that exported the fine material and exhausted the organic matter in the soil.[138] In the Aísa Valley Experimental Station (Spanish Pyrenees), where traditional farming practices and their abandonment were reproduced over 20 years, it was found that plots reproducing shifting agriculture had the highest erosion rates, which generated a large nutrient deficit and slowed plant succession.[139,150,151]

Ruiz Flaño et al.[148] interpreted the greater degradation in some older abandoned fields as a consequence of interactions between plant colonization and land management after abandonment. Regular burning of some

Fig. 11 Abandoned hillslope in the Central Spanish Pyrenees. This hillslope was cultivated with cereal until the first decades of the 20th century through shifting agriculture, favouring soil degradation processes. Almost a century after land abandonment, there is only one cleared shrubland, so soil losses are really high. Photo. T. Lasanta, June 2017.

fields to remove shrubs and facilitate the regeneration of grazing land meant that there was a temporary increase in severe sheet was erosion and some extremely active processes (rills, small gullies). In this respect, Cerdà and Lasanta[152] and Lasanta and Cerdà[153] found that runoff and erosion rates increased after fires, while at the same time, large amounts of important nutrients in the soil were lost. After a few years, the shrubs returned and soil erosion decreased, consequently causing problems for grazing, so the field was burnt again. This repeated process is probably the reason why some fields have highly degraded soils.[93,148] Navas et al.[154] studied the redistribution of Cesium-137 in the soil of a small catchment of abandoned fields, and observed that the sunny hillslopes (most used for crops and, therefore, where shrubs were burnt most frequently) had higher erosion rates and slower plant succession.

Many of the studies on soil erosion in abandoned fields were carried out on terraced slopes, mainly in the Mediterranean basin,[55,96,108,144,155–162] although the problem has also been analyzed in other mountain areas of the world.[163–165]

Terraces are artificial constructions consisting of a dry-stone wall or external enclosure (Fig. 12A) or an earth embankment, frequently with an herbaceous cover (Fig. 12B), to support a flat or slightly sloping area used for crops. They were built to help water infiltration and control soil erosion,

Fig. 12 (A) Abandoned terraces in the Maestrazgo (Iberian System, Spain). They consist in a flat terrace that was cultivated until the middle of the 20th century, and an outer dry-stone wall. In some walls, collapses of the stone walls were observed. Photo. P. Bravo Domínguez. (B) Abandoned terraces in the Central Spanish Pyrenees. In this case, the outer wall is an herbaceous embankment. Stone walls were built in hillslopes with abundant stoniness and/or in sub-mediterranean environments, while in more humid environment the terraces with an herbaceous embankment were more frequent. The good rooting of the grass was enough to maintain the terrace. In addition, it was a less expensive system to construct and maintain (simple cattle grazing). Photo. T. Lasanta, November 2018.

Fig. 13 Abandoned fields on a reverse slope (Iberian System, Spain). To reduce the slope of the fields, the farmer built terraces with an inclined area. Runoff was directed towards the gully that is observed in the hillslope by means of a system of ditches. In the photo, small paths due to livestock grazing were also observed. Photo. R. Lasanta Santolaya, 2014.

and are one of the best methods to prevent soil degradation on cultivated hillslopes.[165,166] Terraces reduce the slope and hydrological connectivity, and control runoff.[55,167,168]

The abandonment of agricultural terraces and the practices involved in their maintenance imply that their capacity to conserve soil is lost.[160,168,169] Several studies have shown that, to prevent erosion and improve the stability of embankments, terraces require constant maintenance both of internal ditches, which mobilize runoff water outside the fields (Fig. 13), and the external enclosures: if, after an intense or lasting rainfall event, a small landslide occurred, the farmer repaired it rapidly and refilled the gap left by the lost soil to prevent the process becoming worse.[55,96,171–176]

In abandoned terraces, collapses and small mass movements are the main erosion process (Fig. 14), forming scars that damage the harmony of the landscape of terraced hillslopes, facilitating gully formation producing high sediment yields. Caneva et al.[172] stated that vegetation, mainly phanerophytes and nano-phanerophytes, growing between the gaps of the stones could facilitate the collapse of walls by putting pressure on the stones, which eventually failed. In addition, the vegetation has chemical effects on the stones, making them disintegrate and break up the terrace.

Fig. 14 Collapses of stone walls in terraces (Iberian System, Spain). These terraces were cultivated with cereal until the 1970s. Nowadays, the outer wall was collapsed in some terraces, and in others it is only partially maintained. On terraced slopes, the highest density of rockslides occurs in the areas with the greatest accumulation of water (concave areas and footslopes) and on very steep slopes, where the height of the terrace is very high, sometimes more than 3 meters as in the photo. Photo. T. Lasanta, April 2019.

Brandolini et al.[161] indicated that the most critical erosion time occurs in the early years after abandoning the terrace, before a dense, structured plant cover has been installed. The same authors observed that young abandoned terraces (less than 25–30 years ago) have erosion rates 2–3 times higher than those abandoned for longer. Revegetation, linked to the age of abandonment, improves the stability of the terraces by reducing the efficacy of morphological processes and steadily decreasing connectivity.[157,173] Moreno-de-las-Heras et al.[162] carried out a meta-analysis of the hydro-geomorphological consequences of the abandonment of agricultural terraces in the Mediterranean region. They concluded that the vegetation cover and the degradation of terraces structures are critical factors for soil degradation in these areas. The authors suggested that the optimum management of terraces, would contribute to the conservation of soil and water resources due to the increase of infiltration rates, decrease in connectivity and the reduction of sediment yield.

Lasanta et al.[55] and Arnáez et al.[174] studied the geomorphological evolution of terraces in Camero Viejo (Iberian Range, Spain). They confirmed that the worst erosion was caused by collapsed of stone walls due to small mass movements and landslides. The highest density of landslides was

Fig. 15 Recent mass movement in a terrace (Cantabrian Range, Spain). This movement is used by cattle in their grazing routes, so in a short time paths will be formed favouring sheet wash erosion. On the other hand, the head of the small mass movement will evolve by headwater erosion, possibly generating a small gully in a few years. In this way, a well-preserved soil will be degraded. Photo. T. Lasanta, April 2019.

observed on concave slopes and the lower part of the hillslopes, coinciding with the higher accumulation of water facilitating saturation of the soil and, therefore, mass movements. Starting at the scar from the landslide of the external wall, the process can progress until forming small gullies. The same authors stated that some of the landslides were originated when livestock passing from one terrace to another (Fig. 15), making paths in which frequent cattle trampling compacts the soil and increases the impact of raindrops and sheet was erosion, and creating scars that are very active. In the same area, García-Ruiz et al.[170] showed that the density of small mass movements in terrace walls had a positive link with the slope gradient and the height of the terrace (higher in steeper hillslopes), and a negative one with the density of plant cover. Lesschen et al.[96] came to similar conclusions in the Cárcavo catchment (Murcia, Spain) and the erosion rate calculated for abandoned terraces is 87 Mg ha^{-1} $year^{-1}$. Brandolini et al.[161] observed in the Vernazza catchment (Italy) that an episode of intense rainfall event recorded in 2011 caused 324 surface landslides, moving 87.000 m^3, a volume comparable with the annual denudation rate registered in small badlands catchments subject to extreme erosion processes.[177] Calsamiglia et al.[111] pointed out that the abandonment of crop terraces changes the spatial patterns of hydrological connectivity, which facilitates concentrated runoff and higher soil erosion and soil degradation. Stavi et al.[168] stated that, on abandoned terraces in the dry lands of the Negev (Southern Israel), during the second half of the 20th century, soil losses rose to 54 Mg km^2 $year^1$ in small hydrological catchments, and to 70 Mg km^2 $year^1$ in large river basins. Soil losses reach 115–180 Mg km^2 year,^{1} with the presence of badland areas in the catchment. Arnáez et al.,[160] based on a review of the hydro-morphological evolution of terraces, indicated that erosion rates differ greatly depending on the dominant processes: erosion rates from runoff are below 3 t ha^{-1} yr^{-1}, whereas they increase if the terrace is affected by mass movements, pipes or gullies (over 100 t ha^{-1} yr^{-1}).

Pallarés Bou[176] observed that, with soil degradation in terraces, subsurface runoff and piping development play an important role, as both processes accelerate mass movements. With heavy or constant rainfall, also in areas with a large accumulation of water, debris flows of more than 100 m can occur and affect all the terraced hillslope. Costra et al.[177] considered that, in these cases, the formation of debris flow is facilitated by the geotechnical and hydraulic properties of the terrain, with the break line located between soil layers with different physical and chemical properties.

Many of the studies on soil erosion in terraces aim to analyze what happens in the inner part of abandoned terraces. Van Dijk et al.[178] indicated that this sector of the terrace is affected by splash and surface runoff. Llorens et al.[155] found that, in the Cal Parisa catchment (eastern Pyrenean, Spain), the inner area is saturated easily during the rainy season, becoming the main source of runoff. This is because the soil is shallow and not very permeable in this part of the terrace, unlike the deeper soil in the outer part. Furthermore, Lasanta et al.[55] observed that in terraces without a flat surface, sheet wash erosion and incisions frequently appear, something also noticed by Romero-Martín et al.[179] in the Guiniguada valley (Gran Canaria, Spain).

In both arid and semi-arid areas, as described above, both in terraces and sloping fields, the extent of soil degradation is closely linked to the availability of water and its effects on the spatial distribution of plant cover and crusts.[98] The age of abandonment in these areas plays a very important role: in the short term, erosion rates tend to be higher than when the terrace was cultivated, since plowing helps infiltration by removing the crusts and increasing the porosity of the soil. In contrast, immediately following abandonment, a crust formed on the surface reduces infiltration and, therefore, causes higher runoff and erosion rates.[78,180] Ries and Hirt[181] concluded that, 5 years after set-aside, crusts occupied 70% of the surface, and led to a reduction in infiltration of at least 50%, which increased runoff and erosion. A few years after abandonment, the level of soil degradation was more affected by the distribution of vegetation and surface crust. Romero-Díaz et al.[113] found that in the Murcia Region (Spain), when crop farming on limestone substrate was abandoned, it led to rapid colonization by herbaceous plants, and later by shrubs, which increased the organic matter in the soil, and was a positive factor in soil conservation and diversity of the landscape. On the other hand, on marly lithology, abandonment produce a surface crust, making plant colonization very slow. At the same time, the soil tends to crack, facilitating channels of subsurface runoff, which facilitate the development of a dense and complex gully network.[13,182]

Ries and Hirt,[181] observed that infiltration generally decreases with the age of abandonment, due to the reduction of porosity from the lack of plowing. Runoff increases due to low infiltration rates but does not lead to high soil loss, since the crust protects it from erosion, at least up to a threshold, after which higher runoff means heavier sheet wash erosion. Lasanta et al.[183] reached the same conclusion in the Ebro Depression (Spain). Cerdà,[78] in a study made in the south-east of Spain, observed that land abandonment in droughts and overgrazing conditions increased soil degradation and reduced infiltration capacity (from 55 mm h^{-1} to 19 mm h^{-1}) and raised erosion rates (from 0 to 334 g m^2 h^{-1}) 3 years after abandonment. Later recovery of vegetation meant that, after 10 years, there was hardly any runoff and soil erosion.

Soil degradation in semi-arid areas is also closely linked to grazing livestock on abandoned fields. Overgrazing increases bulk density and encourage the formation of crusts, reducing infiltration and intensifying erosion. Likewise, grazing livestock can break the crust and create soil particles that are easily transported by wind.[98]

5. Management of abandoned land: restoration and reincorporation into the production system

The results from above highlight the fact that no or poor management is one of the principal causes of soil degradation in abandoned lands. Pardini et al.[120] suggested that the state of degradation in abandoned fields in the Spanish Pyrenees is mainly due to incorrect management. Stavi et al.[168] found that well-preserved terraces had a positive effect on the soil and vegetation, whereas those abandoned and not maintained accelerate soil erosion and land degradation. Lack of management is explained by the fact that land was often abandoned without planning, in answer to the fairly rapid collapse of the society that managed it, or political initiatives, which sought to solve problems in selling agricultural products, rather than managing the land properly.[38] However, the wide area occupied by abandoned land demands management to minimize some of the negative impacts: fire risk, uniform landscape, loss of grazing land, soil degradation, among others.[184] On the other hand, it must be remembered that many abandoned fields still have sufficiently high-quality soil for forestry or agricultural uses, and can be reincorporated into the production system.[121] In addition, climate change dynamics with higher heavy rainfall events, together with the forecast of more land to be abandoned, increases the risk of land degradation.[8] However, such contrasting results on soil degradation suggest that land abandonment should be programmed and managed with soil erosion control strategies to avoid land degradation.[10]

Cogliastro et al.[185] and Benjamin et al.[186] indicated that abandoned lands could be an opportunity for active management establishing new land uses. On the other hand, Stoate et al.[5] favor passive management promoting the use of natural resources and large-scale restoration of the habitats required to recover wildlife. Both alternatives have positive and negative effects on the environment, landscape and socio-economics, and therefore should be discussed between scientists and land managers.[14,81,121,187–189]

Those supporting passive management claim that revegetation after land abandonment leads to the rewilding of ecosystems.[190] Land abandonment is seen as an opportunity to develop self-sufficient natural systems that protect biodiversity and the natural ecosystem processes.[191–193] A more natural landscape can lead to more wild ungulates fulfilling the traditional role of extensive livestock grazing in maintaining a diverse and stable.[194] Navarro and Pereira[14] pointed out that the Mediterranean landscape is more sustainable if the land is more rewilded, since the risk of the erosion processes is reduced. They emphasized that regeneration of the forests could provide certain ecosystem services, such as carbon sequestration, and the creation of leisure areas. The same authors suggested that those wishing to maintain traditional uses, especially agriculture, underestimate the huge amount of work people needs for the sustainability.

The supporters of active management point out that the advantages of rewilding are more apparent than real,[187] and so advocate intervention to avoid the negative effects. They support the control of plant succession leading to an over-rewilded landscape, but also to maintain the traits of a traditional cultural landscape, because of its high ecological,[195] productive,[196,197] and esthetic[198,199] value. In addition, those wishing for intervention say that, to be sustainable, landscapes must be multifunctional: they need to produce food, contribute to social structuring and maintain environmental values. Only in this way they can help in the sustainable development of the population, ensure that the resources and values of the region are upheld and provide ecosystem services to society.[196,200,201] Lastly, some authors believe that abandoning land without active management reduces the quality of the landscape and makes it lose its identity, which could affect the local population psychologically, causing them to feel isolated, poor, ashamed, etc.[63,186]

Active management of abandoned land must take into account the place and concomitant environmental conditions. In arid and semi-arid areas, natural recovery of the landscape is very slow, being its presence and spatial distribution a determining factor in soil degradation. Therefore, a strategy facilitating revegetation must be used. In this respect, Yannelli et al.[180] concluded that management based exclusively on plant succession is not the best strategy for preserving the soil, but practices in the potential recovery of vegetation are needed. Lasanta et al.[183] found that adding fertilizer at the time of abandonment facilitated a rapid increase in the density of plant cover, which improved soil structure and porosity. It also increased infiltration rates and reduced runoff and erosion. On the other hand, the importance of water distribution in vegetation patterns requires management strategies at detailed soil scale, as at this level differences in the quality of soil are seen, soil particles are loosened and runoff begins.[10] In addition, work must be done at hillslope or catchment scale to prevent runoff from connecting with bare areas, which would generate furrows and gullies.[12]

In mountain areas, where revegetation can become an environmental, landscape and socio-economic problem, management after abandonment should include shrub clearing to facilitate extensive livestock grazing and control wildfires. This strategy has been put into practice for more than 30 years in La Rioja (Spain), where approximately 25% of mountain shrubland has been cleared. The clearing programs include maintaining trees and shrubs along ravines and the margins of former fields, conserving a mosaic landscape with cleared areas alternating with others covered in shrubs and/or forest. Moreover, steeper hillslopes (>30%) cannot be cleared, in order to control soil erosion, and trees and shrubs over 2 m tall cannot be cut under any circumstances (Fig. 16). The results are highly positive for the landscape structure,[202] wildfires control[203] and development of extensive grazing livestock.[184]

Terrace conservation and associated maintenance tasks are an essential objective, since their abandonment involves rapid soil degradation, with

Fig. 16 A shrub cleared hillslope (Iberian System, Spain). One of the alternatives to manage abandoned fields is shrub clearing practices. Since 1986, the Regional Government of La Rioja (Spain) has cleared more than 25% of the shrub area in the Iberian System of La Rioja. This practice pursues livestock objectives (regenerate pastures for the development of extensive livestock), environmental (control fires, increase biodiversity) and landscape (improve landscape structure). Photo. T. Lasanta, December 2016.

effects *in situ* and downstream. In addition, climate change predicts more frequent droughts, especially in low and middle latitudes.[204] This makes it more important to maintain the former runoff collection systems, such as terraces.[168] In addition, together with the advantages for the environment, ecosystem services must be taken into account, as they provide services to the population and have a heritage value, and some terraced areas have been recognized by UNESCO as world heritage sites.[6] In that sense, various scientists and land managers have demanded the recovery and sustainable management of the terrace systems.

Occasionally, there are attempts to speed up the recovery of the plant cover by reforestation (Fig. 17), and try to facilitate restoration of soils, improve infiltration conditions, store water, reduce flood peaks and sediment transport.[140,205,206] This strategy is very often used in the Mediterranean countries and it has been seen as a land degradation measure after cropland abandonment, but also recently as a carbon storage practice. In Spain, over 5 million ha were reforested after 1940.[207] Reforesting also has economic objectives (create jobs, produce wood and wood pulp) and territorial (prevent depopulation and restore the landscape). In Spain, for example, the Spanish Forestry Plan was drawn up in 2002 which, among other objectives, proposed to reforest 3,800,000 ha with the consequent

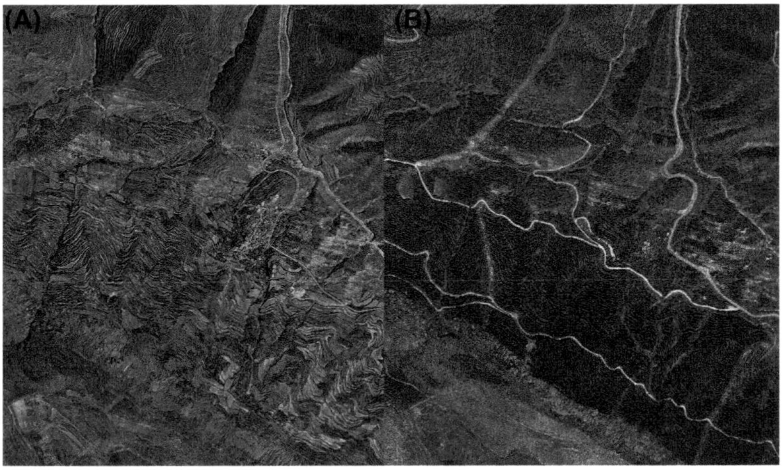

Fig. 17 (A) Cultivated fields and young abandoned fields (1956). (B) Afforested hillslope (2014) (Iberian System, Spain)

capture of up to 60 million tonnes of carbon over the thirty years that the plan would be in force.[208] Different studies, have compared the effects of secondary succession and afforested lands on abandoned agricultural fields. In semi-arid areas, the results suggest that to reach a similar recovery of soil quality and carbon content, longer time is needed in natural succession areas and consequently the soil remains longer prone to degradation (due to low vegetation and soil stability).[209] However, in humid mountain areas soil quality is similar in secondary succession in comparison to afforested areas, and differences in soil organic carbon are limited only to the first 10 cm.[210] Worldwide literature suggests that there are still many uncertainties in the effects of afforestation in soil properties, soil carbon sequestration and soil degradation processes and these topics should be considered in the design of future land abandonment restoration strategies.

6. Conclusions

Soil degradation is nowadays one of the major worldwide environmental problems and land abandonment is one of the main causes. Therefore, this research topic has been studied worldwide, especially since the first decade of this century. A large number of published documents refers to mountain areas, since in these regions land abandonment has been a widespread process. But not only mountain areas have experience agricultural abandonment: (i) land retirement in the European Union due to the CAP, with the objective of regulating the markets of agricultural products, and (ii) the land abandonment in Central and Eastern Europe, in which after 1989 large areas were abandonment in a few years, have been also analyzed.

Land abandonment is a global phenomenon with important implications in soil degradation on several spatial and temporal scales. Land abandonment has been a widespread process in Mediterranean mountains and in arid and semi-arid areas producing active soil degradation processes. Added to this is the fact that abandoned fields are located on very steep hillslopes, in areas with an irregular rainfall regime and frequent heavy rainfall events, and shallow soils.

Studies carried out on the hydro-morphological evolution of abandoned fields show that it is a complex problem, with highly variable results, depending on the field type (terraced, sloping, flat), on the climate conditions that regulate plant succession, on land management during the agricultural stage, which can determine the strength and fertility of the soil at the time of abandonment, and on management after land abandonment, which has an influence through possible disturbances (fires, grazing, etc.). In addition, the age of abandonment plays an important role in the evolution of edaphic properties and plant colonisation.

In abandoned fields on mountain slopes, where rainfall is sufficient to promote plant colonization, the highest soil degradation rates are reached in the first few years following abandonment, when the plant cover is sparse. Once dense shrubland spreads over the whole field, runoff and sediment yielded are reduced to the minimum values. Plant cover improves soil quality which, in turn, facilitates infiltration and water storage. Only if the shrubland is frequently burnt, or if there is overgrazing, plant succession is hampered and highly active erosion processes can be established, leading to very degraded soil, which is sometimes irreversible unless ecological restoration is carried out. Without negative disturbances from humans, abandoned fields on hillslopes turn into slightly eroded areas, with increasingly dense vegetation, high infiltration rates, well-structured soil with a larger content of organic matter.

In abandoned agricultural terraces, erosion is directly related with the density of plant cover, soil properties, environmental conditions and age of abandonment. The most important erosion process is the collapse of outer walls due to small mass movements and landslides. During the cultivated stage, landslides also occurred with heavy or constant rainfall, but farmers quickly repaired the wall and replaced the fallen soil, so soil degradation was rapidly controlled. It must be remembered that terraces were constructed to conserve water and soil, thus allowing crops to be grown. However, when they were abandoned and no longer maintained, they became a key source of sediment export and soil loss that had been accumulated at great effort over centuries. Therefore, terraces are a highly efficient water and soil conservation system, when provided maintenance is kept up.

In arid and semi-arid areas, spatial distribution patterns of vegetation and surface crusts affect infiltration, runoff and soil erosion rates. In general, abandonment brings with it an improvement in soil quality, making it less vulnerable to soil erosion. Nevertheless, the age of abandonment, lithology and any human disturbances play an important role. In chalky or marly soils, it has been observed that vegetation has huge difficulty in progressing, while crusts occupy wide areas. In these cases, rills, gullies and piping are generated (Fig. 18). Among other types of lithology, such as metamorphic substrates, the vegetation gradually covers the fields, improving soil quality and preventing erosion processes from becoming established. Livestock grazing can act by impeding plant succession in some cases, and eliminating crusts in others.

Climate conditions are a clear spatial variability factor in the evolution of abandoned fields and, therefore, in soil degradation. If human intervention does not interrupt plant succession, erosion gradually decreases over time in mountain areas. The process can be very rapid, so that in a few years the land can be protected and the soil quality improved. In semi-arid areas, colonization is much slower, so the soil gradually degrades and highly aggressive erosion processes are established.

Fig. 18 Collapses in gypsums associated to piping processes. Photo. T. Lasanta.

This chapter also emphasizes the need to program land abandonment and properly manage those lands that have already been abandoned, with soil conservation programs that, as far as possible, are compatible with sustainable rural development and the creation of mosaic landscapes. In this respect, extensive livestock grazing seems to be a good strategy to mitigate soil degradation due to land abandonment and support the development of marginalized areas.

Acknowledgments

This research was supported by the ESPAS project (CGL2015-65569-R, funded by the MINECO-FEDER). The "Geomorphology and Global Change" (E02_17E) research group was financed by the Aragón Government and the European Social Fund (ESF-FSE).

References

1. Kosmas C, Kairis O, Karavitis C, Ritsema C, Salvati L, Acikalin S, Alcalá M, Alfama P, et al. Evaluation and selection of indicators for land degradation and desertification monitoring: methodological approach. *Environ Manage* 2004;**54**:951–70.
2. Smiraglia D, Ceccarelli T, Bajocco S, Salvati L, Perini L. Linking trajectories of land change, land degradation processes and ecosystem services. *Environ Res* 2016;**147**: 590–600.
3. Le Houérou HN. Agroforestry and sylvopastoralism to combat land degradation in the Mediterranean basin: old approaches to new problems. *Agric Ecosyst Environ* 1990;**33**: 99–109.
4. EC. *European Commission – towards a Thematic strategy for soil protection.* 2002 (COM (2002) 179 final).
5. Stoate C, Báldi A, Beja P, Boatman ND, Herzon I, van Doorn A, de Snoo GR, Rakosy L, Ramwell C. Ecological impacts of early 21st century agricultural change in Europe— a review. *J Environ Manag* 2009;**91**:22–46.
6. Romero-Díaz A, de Vente J, Díaz-Pereira E. Assessment of the ecosystems services provided by agricultural terraces. *Pirineos* 2019:e043.
7. Geist HJ, Lambin EF. Dynamic causal patterns of desertification. *Bioscience* 2004;**54**: 817–29.
8. Bajocco S, De Angelis A, Perini L, Ferrara A, Salvati L. The impact of land use/land cover change on land degradation dynamics: a Mediterranean case study. *Environ Manage* 2012;**49**:980–9.
9. Vergari F, Della Seta M, Del Monte M, Fredi P, Lupia PE. Long-and short-term evolution of several Mediterranean denudation hot spots: the role of rainfall variations and human impact. *Geomorphology* 2013;**183**:14–27.
10. Rodrigo-Comino J, Martínez-Hernández C, Iserloh T, Cerdà A. Contrasted impact of land abandonment on soil erosion in Mediterranean agriculture fields. *Pedosphere* 2018; **28**(4):617–31.
11. Thornes JB. Mediterranean desertification and the vegetation cover. In: Frantechi R, Peter D, Balabanis P, Rubio JL, editors. *Desertification in a European context: physical and socio-economic aspects.* Luxembourg: Office for Official Publications of the European Communities; 1995. 169–194pp.
12. Cammeraat LH, Imeson AC. The evolution and significance of soil-vegetation patterns following land abandonment and fire in Spain. *Catena* 1999;**37**:107–27.

13. Lesschen JP, Kok K, Verburg PH, Cammeraat LH. Identification of vulnerable areas for gully erosion under different scenarios of land abandonment in Southeast Spain. *Catena* 2007;**71**:110–21.
14. Navarro LM, Pereira HM. Rewilding abandoned landscapes in Europe. *Ecosystems* 2012;**15**:900–12.
15. Brevik EC, Cerdà A, Mataix-Solera J, Pereg L, Quinton JN, Six J, Van Oost K. The interdisciplinary natute of SOIL. *SOIL* 2015;**1**:117–9.
16. Ramankutty N, Foley JA. Estimating historical changes in global land cover: croplands from 1700 to 1992. *Glob Biogeochem Cycles* 1999;**13**:997–1027.
17. Pointereau P, Coulon F, Girard P, Lambotte M, Stuczynski T, Sánchez Ortega V, Del Rio A. *Analysis of farmland abandonment and the extent and location of agricultural areas that are actually abandoned or are in risk to be abandoned*. JRC Scientific and Technical; 2008. Reports (EUR 23411) In: http://publicationsjrceceuropaeu/repository/handle/111111111/166.
18. Strijker D. Marginal lands in Europe —causes of decline. *Basic Appl Ecol* 2005;**6**:99–106.
19. Kosmas C, Kairis O, Karavitis C, Acikalin S, Alcalá M, Alfama P, Atlhopheng J, et al. An exploratory analysis of land abandonment drivers in areas prone to desertification. *Catena* 2015;**128**:252–61.
20. Baudry J. Ecological consequences of grazing extensification and land abandonment: role of interactions between environment, society and techniques. *Options Méditerranéennes* – Serie Séminaires 1991;**15**:13–9.
21. Pinto-Correia T. Land abandonment: changes in the land use patterns around the Mediterranean basin. *Cah Options Mediterr* 1993;**1**(2):97–112.
22. Martínez-Hernández C. *Metodología de evaluación y análisis preliminar de la superficie en abandono agrícola de la Región de Murcia*. PhD Dissertation. Murcia. 2014.
23. Romero-Díaz A, Coord. *Abandono de cultivos en la región de Murcia*. Murcia: Consecuencias ecogeomorfológicas. Edit.um; 2016. p. 263.
24. Alonso-Sarriá F, Martínez-Hernández C, Romero-Díaz A, Cánovas-García F, Gomariz-Castillo F. Main environmental features leading to recent land abandonment in Murcia Region (Southeast Spain). *Land Degrad Dev* 2016;**27**:654–70.
25. Keenleyside C, Tucker GM. *Farmland abandonment in the EU: an assessment of trends and prospects*. London: Report prepared for WWF. Institute for European Environmental Policy; 2010. p. 93.
26. García-Ruiz JM, López-Bermúdez F. *La erosión del suelo en España*. Zaragoza: Sociedad Española de Geomorfología; 2009. p. 441.
27. Fetcher W, Lesslar JE, Lewthwaite R. The aetiology of the tsutsugamushi disease and tropical typhus in the Federated Malay state. *Trans R Soc Trop Med Hyg* 1928;**22**(2):161–2.
28. Fuller GL. Charting the effects of erosion in the old plantation belt of the southern Piedmont. Eos. *Trans Am Geophys Union* 1934;**15**(2):495–500.
29. Egler FE. Vegetation science concepts I. Initial floristic composition, a factor in old field vegetation development with 2 figs. *Vegetatio Acta Geobotanica* 1954;**4**(6):412–7.
30. Hill AR. Factors affecting the export of nitrate-nitrogen from drainage basins in southern Ontario. *Water Res* 1978;**12**:1045–57.
31. Stringer LC, Harris A. Land degradation in Dolj county, southern Romania: environmental Changes, impacts and responses. *Land Degrad Dev* 2014;**25**:17–28.
32. Hill J, Stellmes M, Udelhoven T, Röder A, Sommer S. Mediterranean desertification and land degradation. Mapping related land use change syndromes based on satellite observations. *Glob Planet Chang* 2008;**64**:146–57.

33. Weissteiner CJ, Borchetti M, Böttcher K, Carrara P, Bordogna G, Brivio PA. Spatial explicit assessment of rural land abandonment in the Mediterranean area. *Glob Planet Chang* 2011;**79**:20–36.
34. Lasanta T. *El paisaje de campos abandonados en Cameros Viejo (Sistema Ibérico, La Rioja)*, vol. 32. Logroño: Instituto de Estudios Riojanos, Ciencias de la Tierras; 2014. p. 305.
35. Antrop M. Changing patterns in the urbanized of western Europe. *Landsc Ecol* 2000; **15**(3):257–70.
36. Serra P, Vera A, Tulla AF, Salvati L. Beyond the urban-rural gradient: intensification and extensification land-use and socioeconomic processes in Spain. *Appl Geogr* 2014; **55**:71–81.
37. MacDonald D, Crabtree JR, Wiesinger G, Dax T, Stamou N, Fleury P, Gutierrez Lazpita J, Gibon A. Agricultural abandonment in mountain areas of Europe: environmental consequences and policy response. *J Environ Manag* 2000;**59**:47–69.
38. García-Ruiz JM, Lana-Renault N. Hydrological and erosive consequences of farmland abandonment in Europe, with special reference to the Mediterranean region – a review. *Agric Ecosyst Environ* 2011;**140**:317–38.
39. Taillefumier F, Piégay H. Contemporary land use changes in Prealpine Mediterranean mountains: a multivariate GIS-based approach applied two municipalities in the Southern French Prealps. *Catena* 2003;**51**:267–96.
40. Debussche M, Lepart J, Desvieux A. Mediterranean landscape changes: evidence from old postcards. *Glob Ecol Biogeogr* 1999;**8**:3–15.
41. Chauchard S, Carcaillet C, Guibal F. Patterns of land-use abandonment control tree-recruitment and forest dynamics in Mediterranean mountains. *Ecosystems* 2007;**10**: 936–48.
42. Walther P. Land abandonment in the Swiss Alps. A new understanding of a land-use problem. *Mt Res Dev* 1986;**6**(4):305–14.
43. Waisamen PJ, Bliss NB. Changes in population and agricultural land in conterminous United States counties, 1790–1997. *Glob Biogeochem Cycles* 2002;**16**:1–19.
44. Mallet M. Agriculture et tourisme dans un milieu haute-alpin: un exemple briançonnais. *Études Rural* 1978;**71–72**:111–54.
45. Réparaz AG. Le recul de l'espace agro-pastoral dans les Prealpes du Sud et ses conséquences sur le paysage de la moyenne montagne. *Pirineos* 1992;**139**:47–66.
46. Darbellay C. Mountain agriculture in change. In: Brugger E, Furrer G, Messerli P, editors. *The transformation of Swiss mountain regions*. Berna; 1984. p. 289–316.
47. Haefner H, Günter T. Land-use changes and ecological effects in the Swiss Alps. In: Brugger E, Furrer G, Messerli P, editors. *The transformation of Swiss mountain regions*. Berna; 1984. p. 101–24.
48. Balent G, Barrué-Pastor M. Pratiques pastorales et strategies foncieres en vallée d'Oô (Pyrénées centrales). *Rev Geogr Pyrenees Sud-Ouest* 1986;**57**(3):403–47.
49. Kozak J, Ostapowicz K, Szablowska-Midor A, Widacki W. Land abandonment in the western Beskidy Mts and its environmental background. *Ekologia Bratislava* 2004;**23**(1): 116–26.
50. Kuemmerle T, Hostert P, Radeloff VC, van der Linden S, Perzanowski K, Kruhlov I. Cross–border comparison of post-socialist farmland abandonment in the Carpathians. *Ecosystems* 2008;**11**(4):614–28.
51. Tasser E, Walde J, Tappeiner U, Teutsch A, Noggler W. Land-use changes and natural reforestation in the eastern central Alps. *Agric Ecosyst Environ* 2007;**118**:115–29.
52. Hostert P. The Carpathian mountains in transformation. *Mountain Research Initiative Newsletter* 2010;**5**:6–8.
53. Lasanta T. El proceso de marginación de tierras en España. In: Lasanta T, García-Ruiz JM, editors. *Erosión y recuperación de tierras en áreas marginales*. Logroño: Instituto de Estudios Riojanos – Sociedad Española de Geomorfología; 1996. p. 7–31.

54. Lasanta T. The process of desertion of cultivated areas in the Central Spanish Pyrenees. *Pirineos* 1988;**132**:15–36.
55. Lasanta T, Arnáez J, Oserin M, Ortigosa L. Marginal lands and erosion in terraced fields in the Mediterranean mountains: a case study in the Camero Viejo (Northwestern Iberian System, Spain). *Mt Res Dev* 2001;**21**(1):69–76.
56. Douglas TD, Kirkby SJ, Critchley RW, Park GJ. Agricultural terrace abandonment in the Alpujarra, Andalucía, Spain. *Land Degrad Rehabil* 1994;**5**:281–91.
57. Ervin D. *Some lessons about the political economic effects of set-aside: the United State's experience.* USA: Department of Agricultural and Resource Economics, Oregon State University; 1992.
58. Ansell DJ, Tranter RB. *Set-aside: in theory and in practice.* University of Reading, Department of Agricultural Economics and Management. Miscellaneous study (University of Reading. Department of Agricultural Economics and Management), no. 75; 1992.
59. Feranec J, Jaffrain G, Soukup T, Hazeu G. Determining changes and flows in European landscapes 1990–2000 using CORINE land cover data. *Appl Geogr* 2010;**30**(1):19–35.
60. IEEP A. *Reflecting environmental land use needs into EU policy: preserving and enhancing the environmental benefits of "land services": soil sealing, biodiversity corridors, intensification/marginalization of land use and permanent grassland.* Final report to the European Commission, DG Environment on Contract ENVB1/ETU/2008/0030 Institute for European Environmental Policy/AlterraWageningen UR. 2010.
61. Verburg PH, Overmars KP. Combining top-down and bottom-up dynamics in land use modeling: exploring the future of abandoned farmlands in Europe with the Dyna-CLUE model. *Landsc Ecol* 2009;**24**(9):1167–81.
62. Nikodemus O, Bell S, Grine I, Liepins I. The impact of economic, social and political factors on the landscape structure of the Vidzeme Uplands in Latvia. *Landsc Urban Plan* 2005;**70**(1–2):57–67.
63. Palang H, Printsmann A, Konkoly Gyuro E, Urbanc M, Skowrenerk E, Woloszyn W. The forgotten rural landscapes of Central and Eastern Europe. *Landsc Ecol* 2006;**21**(3):347–57.
64. Melluma A. Metaporphoses of Lavian landscapes during fifty years of Soviet rule. *Geojournal* 1994;**33**(1):55–62.
65. Bell S, Mortarzino O, Aspinall P, Peneze Z, Nikodemus O. Rural society, social inclusion and landscape change in Central and Eastern Europe: a case study of Latvia. *Sociol Rural* 2009;**49**(3):295–326.
66. Bell S, Alves S, de Oliveira S, Zuin EA. Migration and land use change in Europe: a review. *Living Reviews of Landscape Research* 2010;**34**(4):425–55.
67. Pasakarnis G, Maliene V. *The land consolidation in central and eastern Europe: redefining the agricultural destiny.* BEAN Conference 2009, BEST Research Centre (Built Environment and Sustainable Technologies). UK: Liverpool John Moores University; 2009.
68. Land abandonment, biodiversity and the CAP. In: Van Dijk GA, Zelanowicz A, Blokziljl R, editors. *Land abandonment and biodiversity in relation to the 1st and 2nd pillars of the EU's Common Agricultural Policy: outcome of an international seminar in Sigulda, Latvia, 7–8 October, 2004. DLC, Government Service for Land and water management, Utrecht, The Netherlands*; 2005.
69. Muller D, Kuemmerle T. *Causes of cropland abandonment during the post-socialist transition in Southern Romania, Regional aspects of climate-terrestrial hydrologic interactions in non-boreal Eastern Europe. NATO Science for Peace and Security Series*, vol. 4; 2009. p. 221–30.
70. Peterson U, Aunap R. Changes in agricultural land use in Estonia in the 1990s detected with multitemporal Landsat MSS imagery. *Landsc Urban Plan* 1998;**41**:193–201.

71. Ruskule A, Nikodemus O, Kasparinskis R, Bell S, Urtane I. The perception of abandonment farmland by local people and experts: landscape value and perspectives on future land use. *Landsc Urban Plan* 2013;**115**:49–61.
72. Nowicki PL, Kneirim A, Banse MAH, Belling M, Helming J, Leibert T, Lentz S, Margraf O, et al. *Scenar 2020: scenario study on agriculture and the rural world. Contract No. 30 – CE – 0040087/00-08 for Directorate-General Agriculture and Rural Development.* 2007.
73. Nowicki P, Goba V, Knierim A, van Meijl H, Banse M, Delbaere B, Helming J, Hunke P, et al. *Scenar 2020-II – update of analysis of prospects in the scenar 2020 study – contract No 30–CE-0200286/00-21 European Commission, Directorate-General Agriculture and Rural Development, Brussels.* 2009.
74. Rienks WA, editor. *The future of rural Europe: an anthology based on the results of the Eururalis study.* Wageningen University Research and Netherlands Environmental Assessment Agency; 2008.
75. Paracchini ML, Terres JM, Petersen JE, Hoogeveen Y. High nature value farmland and traditional agricultural landscapes. In: Pedroli B, van Doorn A, de Blust G, Paracchini ML, Wascher D, Bunce F, editors. *Europe's living landscapes. Essays on Exploring our identity in the countryside.* Wageningen: Landscape Europe/Knnv; 2007. p. 22–34.
76. Nainggolan D, de Vente J, Boix-Fayos C, Termansen M, Hubacek K, Reed MS. Afforestation, agricultural abandonment and intensification: Competing trajectories in semi-arid Mediterranean agro-ecosystems. *Agric Ecosyst Environ* 2012;**159**:90–104.
77. Thornes JB. The ecology of erosion. *Geography* 1985;**70**:222–36.
78. Cerdà A. Soil erosion after land abandonment in a semiarid environment of southeastern Spain. *Arid Land Res Manag* 1997;**11**(2):163–76.
79. Pugnaire FI, Luque MT, Armas C, Gutiérrez L. Colonization processes in semiarid Mediterranean old-fields. *J Arid Environ* 2006;**65**:591–603.
80. Mottet A, Ladet S, Coque N, Gibón A. Agricultural land-use change and its drivers in mountain landscapes: a case study in the Pyrenees. *Agric Ecosyst Environ* 2006;**114**:296–310.
81. Gellrich H, Baur P, Koch B, Zimmermann NE. Agricultural land abandonment and natural forest re-growth in the Swiss mountains: a spatial explicit economic analysis. *Agric Ecosyst Environ* 2007;**118**:93–108.
82. Teira AG, Peco B. Modelling old field species richness in a mountain area. *Plant Ecol* 2003;**166**(2):249–61.
83. Pedroli B, Pinto-Correia T, Cornish P. Landscape –what's in it? Trends in European landscape science and priority themes for concerted research. *Landsc Ecol* 2006;**21**: 421–30.
84. Zhao HL, Zhao XY, Zhou RL, Zhang TH, Drake S. Desertification processes due to heavy grazing in Sandy rangeland, Inner Mongolia. *J Arid Environ* 2005;**62**:309–19.
85. Mitchley J, Ispikoudis I. Grassland and shrubland in Europe: biodiversity and conservation. *Grassland Sci Europe* 1999;**4**:239–51.
86. Zaravali MP, Yakoulaki MD, Papansatasis VP. Effects of shrub encroachment on herbage production and nutritive value in semi-arid Mediterranean grasslands. *Grass Forage Sci* 2007;**62**:355–63.
87. Sluiter R, De Jong SM. Spatial patterns of Mediterranean land abandonment and related land cover transition. *Landsc Ecol* 2007;**22**:559–76.
88. Tzanopoulos J, Mitchley J, Pantis JD. Vegetation dynamics in abandoned crop fields on a Mediterranean island: development of succession model and estimation of disturbance thresholds. *Agric Ecosyst Environ* 2007;**83**:83–94.
89. Molinillo M, Lasanta T, García-Ruiz JM. Managing degraded landscape after farmland abandonment in the Central Spanish Pyrenees. *Environ Manage* 1997;**21**:587–98.

90. San Miguel-Ayanz J, Moreno JM, Camia A. Analysis of large fires in European Mediterranean landscapes: lessons learned and perspectives. *Ecol Manag* 2013;**249**:11–22.
91. San Emeterio L, Múgica L, Ugarte MD, Goicoa T, Canals RM. Sustainability of traditional pastoral fires in highlands under change: effects on soil function and nutrient cycling. *Agric Ecosyst Environ* 2016;**235**:155–63.
92. Nunes A, Coelho C, Almeida A, Rodríguez A. Soil erosion and hydrological response to land abandonment in a central inland area of Portugal. *Land Degrad Dev* 2010;**21**: 260–73.
93. Lasanta T, Nadal-Romero E, Errea MP. The footprint of marginal agriculture in the Mediterranean mountain landscape: an analysis of the Central Spanish Pyrenees. *Sci Total Environ* 2017;**599–600**:1823–36.
94. Pueyo Y, Beguería S. Modelling the rate of secondary succession after farmland abandonment in a Mediterranean mountain area. *Landsc Urban Plan* 2007;**83**:245–54.
95. Acácio V, Holmgren M, Jansen PA, Schrotter O. Multiple recruitment limitation causes arrested succession in Mediterranean cork oak systems. *Ecosystems* 2007;**10**: 1220–30.
96. Lesschen JP, Cammeraat LH, Nieman T. Erosion and terrace failure due to agricultural land abandonment in a semi-arid environment. *Earth Surf Process Landf* 2008;**33**:1574–84.
97. Puigdefábregas J. The role of vegetation patterns in structuring runoff and sediment fluxes in dryland. *Earth Surf Process Landf* 2005;**30**:133–47.
98. Ries JB, Langer M, Rehberg C. Experimental investigations on water and wind erosion on abandoned fields and arable land in the central Ebro Basin, Aragon/ Spain. *Z Geomorphol* 2000;**121**:91–108.
99. Haase P, Pugnaire FI, Clark SC, Incoll LD. Spatial pattern in *Anthyllis cytisoides* shrublands on abandoned land in southeastern Spain. *J Veg Sci* 1997;**8**:627–34.
100. Ruskule A, Nikodemus O, Kasparinskis R, Prizavoite D, Bojare D, Brumelis G. Soil-vegetation in abandoned farmland within the temperate region of Europe. *New For* 2016;**47**:587–605.
101. Bonet A, Pausas J. Species richness and cover along a 60-year chronosequence in old-fields of southeastern Spain. *Plant Ecol* 2004;**174**:257–70.
102. Rey-Benayas JM, Martínez Baroja L, Pérez-Camacho L, Villar-Salvador P, Holl KD. Predation and aridity slow down the spread of 21-year-old planted Woodland islets in restored Mediterranean farmland. *New For* 2015;**46**:841–53.
103. Verdú M, García-Fayos P. Old-field colonization by *Daphne gnidium*: distribution and spatial dependence at different scales. *J Veg Sci* 1998;**9**:713–8.
104. Clements FE. *Plant succession and indicators*. New York: Wilson; 1928. p. 453.
105. Clements FE. Nature and structure of the climax. *J Ecol* 1936;**24**:252–84.
106. Ruskule A, Nikodemus O, Kasparinskis R, Brümelis G. Patterns of afforestation on abandoned agriculture land in Latvia. *Agrofor Syst* 2012;**85**(2):215–31.
107. Peña-Angulo D, Khorchani M, Errea P, Lasanta T, Martínez-Arnáiz M, Nadal-Romero E. Factors explaining the diversity of land cover in abandoned fields in a Mediterranean mountain area. *Catena* 2019;**181**:104064.
108. Koulouri M, Giurga CH. Land abandonment and slope gradient as key factors of soil erosion in Mediterranean terraced lands. *Catena* 2007;**69**:274–81.
109. Arnáez J, Lasanta T, Errea MP, Ortigosa L. Land abandonment, landscape evolution, and soil erosion in a Spanish Mediterranean region. The case of Camero Viejo. *Land Degrad Dev* 2011;**22**:537–50.
110. Belmonte Serrato F, Martínez Hernández C, Ruiz Sinoga JD. Análisis de las características edáficas más significativas de las áreas abandonadas. In: Romero AD, (Coord.), editors. *Abandono de cultivos en la región de Murcia. Consecuencias ecogeomorfológicas*; 2016. p. 41–52. Murcia.

111. Calsamiglia A, Lucas-Borja ME, Fortesa J, García-Comendador J, Estrany J. Changes in soil quality and hydrological connectivity caused by the abandonment of terraces in a Mediterranean burned catchment. *Forests* 2017;**8**:333.
112. Lasanta T, Vicente-Serrano SM, Cuadrat JM. Mountain Mediterranean landscape evolution caused by the abandonment of traditional primary activities: a study of the Spanish Central Pyrenees. *Appl Geogr* 2005;**25**:47–65.
113. Romero-Díaz A, Marín Sanleandro F, Sánchez Soriano M, Belmonte Serrato F, Faulkner H. The causes of piping in a set of abandoned agricultural terraces in Mediterranean Spain. *Land Degrad Dev* 2007;**9**:179–88.
114. Romero-Díaz A, Ruiz-Sinoga JD, Robledano-Aymerich F, Brevik E, Cerdá A. Ecosystem responses to land abandonment in western Mediterranean mountains. *Catena* 2017;**149**:824–35.
115. Pardini G, Gispert M. Soil quality assessment through a multi-approach analysis in soils of abandoned terraced land in NE Spain. *Cuadernos de Investigación Geográfica* 2012; **38**(2):7–30.
116. de Baets S, Meersmans J, Vanacker V, Quine TA, Van Oost K. Spatial variability and change in soil organic carbon stocks in response to recovery following land abandonment and erosion in mountainous drylands. *Soil Use Manag* 2013;**29**(1):65–76.
117. Van Hall RL, Cammeraat LH, Keesstra SD, Zorn M. Impact of secondary vegetation succession on soil quality in a humid Mediterranean landscape. *Catena* 2017;**149**:836–43.
118. Bautista Cruz A, Etchevers Barra J, del Castillo RF, Gutiérrez C. La calidad del suelo y sus indicadores. *Ecosistemas* 2004;**13**:90–7.
119. Pardini G, Gispert M. Impact of land abandonment on water erosion in soils of the Eastern Iberian Peninsula. *Agrochimica* 2006;**50**(1–2):13–24.
120. Pardini G, Aringheri R, Plana F, Gallart F. Soil properties relevant to land degradation in abandoned sloping fields in Aísa Valley, Central Pyrenees (Spain). *Pirineos* 1991;**137**: 79–93.
121. Lasanta T, Nadal-Romero E, Arnáez J. Managing abandoned farmland to control the impact of re-vegetation on the environment. The state of art in Europe. *Environ Sci Policy* 2015;**52**:99–109.
122. Ruiz-Sinoga JD, Martínez-Murillo JF. Hydrological response of abandoned agricultural soils along a climatological gradient on metamorphic parent material in southern Spain. *Earth Surf Process Landf* 2009;**34**:2047–56.
123. Rietkert M, Dekker SC, de Ruiter PC, van de Koppel J. Self-organized patchiness and catastrophic shifts in ecosystems. *Science* 2004;**305**:1926–9.
124. Segoli M, Ungar ED, Shachak M. Shrubs enhance resilience of a semi-arid ecosystem by engineering and regrowth. *Ecohydrology* 2008;**1**:330–9.
125. Moret D, Pueyo Y, Bueno CG, Alados CL. Hydro-physical responses of gypseous and non gypseous soils to livestock grazing in a semi-arid region of NE Spain. *Agric Water Manag* 2011;**98**:1822–7.
126. Pueyo Y. Aportación a los modelos ecohidrológicos con *feedbacks* al conocimiento del funcionamiento de los ecosistemas de zonas áridas y semi-áridas. *Cuadernos de Investigación Geográfica* 2013;**39**(2):243–58.
127. Mills A, Fey M, Donaldson J, Todd S, Theron L. Soil infiltration as a driver of plant cover and species richness in the semi-arid Karoo, South Africa. *Plant Soil* 2009;**320**: 321–32.
128. Thornes JB. Modelling soil erosion by grazing. Recent developments and new approaches. *Geogr Res* 2007;**45**(1):13–26.
129. Turnbull L, Wilcox BP, Belnap J, Ravi S, D'Orico P, Childers D, Gwenzi W, Okin G, Wainwright J, Caylor KK, Sankey T. Understanding the role of ecohydrological feedbacks in ecosystem state change in drylands. *Ecohydrology* 2012;**5**:174–83.

130. Pueyo Y, Alados CL, García-Ávila B, Kefi S, Maestro M, Rietberk M. Comparing direct abiotic amelioration and facilitation as tools for restoration on semi-arid grassland. *Restor Ecol* 2009;**17**:908–16.
131. Wertabach TM, Hazel N, Kampf I, Yurtaev A, Tupitsin S, Kiehl K, Kamp J, Kleinebecker T. Soil carbon sequestration due to post-Soviet cropland abandonment: Estimates from a large-scale soil organic carbon field inventory. *Glob Chang Biol* 2017; **23**:3729–41.
132. Tommaso C, Emanuele B, Guido P, Lucia P, Vincenza CM, Riccardo V. Soil organic carbon pool's contribution to climate mitigation on marginal land of a Mediterranean montane area in Italy. *J Environ Manag* 2017;**218**:593–601.
133. Gabarrón-Galeote MA, Trigalet S, van Wesemael B. Soil organic carbon evolution after land abandonment along a precipitation gradient in southern Spain. *Agric Ecosyst Environ* 2015;**199**:114–23.
134. Novara A, Gristina L, Sala G, Galati A, Crescimanno M, Cerdà A, Badalamenti E, La Mantia T. Agricultural land abandonment in Mediterranean environment provides ecosystem services via soil carbon sequestration. *Sci Total Environ* 2017;**756**:420–9.
135. Gosheva S, Walthert L, Niklaus PA, Zimmermann S, Gimmi U, Hagedorn F. Reconstruction of historic Forest cover changes indicates minor effects on carbon stocks in Swiss Forest soils. *Ecosystems* 2017;**20**(8):1512–28.
136. Nadal-Romero E, Pereira P, Martínez-Murillo JF. Editorial: soil processes in mountain environments. *Sci Total Environ* 2019;**656**:701–8.
137. Minasny B, Malone BP, McBratney AB, Angers DA, Arrouays D, Chambers A, et al. Soil carbon 4 per mille. *Geoderma* 2017;**292**:59–86.
138. Seeger M, Ries JB. Soil degradation and soil surface process intensities on abandoned fields in Mediterranean mountain environments. *Land Degrad Dev* 2008;**19**:488–501.
139. Lasanta T, Beguería S, García-Ruiz JM. Geomorphic and hydrological effects of traditional shifting agriculture a Mediterranean mountain, Central Spanish Pyrenees. *Mt Res Dev* 2006;**26**(2):146–52.
140. García-Ruiz JM. The effects of land uses on soil erosion in Spain: a review. *Catena* 2010;**81**:1–11.
141. Martínez-Fernández J, López-Bermúdez F, Martínez-Fernández J, Romero-Díaz MA. Land use and soil-vegetation relationships in a Mediterranean ecosystem. *Catena* 1995; **25**:153–67.
142. Kosmas C, Danalatos NG, Gerontidis ST. The effect of land parameters on vegetation performance and degree of erosion under Mediterranean conditions. *Catena* 2000;**40**: 3–17.
143. López-Bermúdez F, Romero-Díaz MA, Martínez-Fernández J, Martínez-Fernández J. The Ardal fieldsite: soil and vegetation cover. In: Brandt J, Thornes JB, editors. *Mediterranean desertification and land use*. Chichester: Wiley; 1996. p. 169–88.
144. Ruecker G, Schad P, Alcubilla MM, Ferrer C. Natural regeneration of degraded soils and site changes on abandoned agricultural terraces in Mediterranean Spain. *Land Degrad Dev* 1998;**9**:179–88.
145. Nadal J, Pélachs A, Molina D, Soriano JM. Soil fertility evolution and landscape dynamics in a Mediterranean area: a case study in the Sant Llorenc Park (Barcelona, NE Spain). *Area* 2009;**41**(2):121–38.
146. Dunjó G, Pardini G, Gispert M. The role of land use-land cover on runoff generation and sediment yield at a microplot scale, in a small Mediterranean catchment. *J Arid Environ* 2004;**57**:99–116.
147. Cerdà A, Rodrigo-Comino J, Novara A, Brevik E, Reza A, Pulido M, Giménez-Morera A, Keestra S. Long-term impact of rainfed agricultural land abandonment on soil erosion in the Western Mediterranean basin. *Prog Phys Geogr* 2018;**42**(2): 202–19.

148. Ruiz-Flaño P, García-Ruiz JM, Ortigosa L. Geomorphological evolution of farmland abandonment fields. A case study in the Central Pyrenees. *Catena* 1992;**19**:301–8.
149. Poesen J, De Luna E, Franca A, Nachtergaele J, Govers G. Concentrated flow erosion rates as affected by rock fragment cover and initial soil moisture content. *Catena* 1999; **36**:323–35.
150. Nadal-Romero E, Lasanta T, García-Ruiz JM. Runoff and sediment yield from land under various uses in a Mediterranean mountain area: long-term results from an experimental station. *Earth Surf Process Landf* 2013;**38**:346–55.
151. Nadal-Romero E, Khorchani M, Lasanta T, García-Ruiz JM. Runoff and solute outputs under different land uses: long term results from a Mediterranean mountain experimental station. *Water* 2019;**11**(5):976.
152. Cerdà A, Lasanta T. Long-term erosional responses after fire in the Central Spanish Pyrenees. 1. Water and sediment yield. *Catena* 2005;**60**:58–80.
153. Lasanta T, Cerdà A. Long-term erosional responses after fire in the Central Spanish Pyrenees. 2. Solute release. *Catena* 2005;**60**:81–100.
154. Navas A, Machín J, Soto J. Assessing soil erosion in a Pyrenean mountain catchment using GIS and fallout 137Cs. *Agric Ecosyst Environ* 2005;**105**:493–506.
155. Llorens P, Latron J, Gallart F. Analysis of the role of agricultural abandoned terraces on the hydrology and sediment dynamics in a small mountainous basin. *Pirineos* 1992;**139**: 27–46.
156. Duiker SW, Flanagan DC, Lal R. Erodibility and infiltration characteristics of five major soils of Southwest Spain. *Catena* 2001;**45**:103–21.
157. Cammeraat E, van Breek R, Kooijman A. Vegetation succession and its consequences for slope stability in SE Spain. *Plant Soil* 2005;**278**:135–47.
158. Hammad AA, Borresen T. Socioeconomic factors affecting farmers' perceptions of land degradation and stonewall terraces in central Palestine. *Environ Manag* 2006;**37**:380–94.
159. Petanidou T, Kizos T, Soulakellis N. Socioeconomic dimensions of changes in the agricultural landscape of the Mediterranean basin: a case of study of the abandonment of cultivation terraces on Nisyros Island, Greece. *Environ Manag* 2008;**41**:250–66.
160. Arnáez J, Lana-Renault N, Lasanta T, Ruiz-Flaño P, Castroviejo J. Effects of farming terraces on hydrological and geomorphological processes. A review. *Catena* 2015;**128**: 122–34.
161. Brandolini P, Cevasco A, Capolongo D, Pepe G, Lovergine F, Del Monte M. Response of terraced slopes to a very intense rainfall event and relationships with land abandonment: a case study from Cinque Terre (Italy). *Land Degrad Dev* 2018; **29**:630–42.
162. Moreno-de-las-Heras M, Linderberger F, Latron J, Lana-Renault N, Llorens P, Arnáez J, Romero-Díaz A, Gallart F. Hydro-geomorphological consequences of the abandonment of agricultural terraces in the Mediterranean region: key controlling factors and landscape stability patterns. *Geomorphology* 2019;**333**:73–91.
163. Vagen TG, Tilahnm Y, Esser KB. Effects of stone terracing on available phosphorus and yields on highly eroded in Tigray, Ethiopia. *J Sustain Agric* 1999;**15**:61–74.
164. Inbar M, Llerena CA. Erosion processes in high mountain agricultural terraces in Peru. *Mt Res Dev* 2000;**20**:72–9.
165. Li XH, Yang J, Zhao CY, Wang B. Runoff and sediment from orchard terraces in southeastern China. *Land Degrad Dev* 2014;**25**:184–92.
166. Louwagie G, Gray SH, Sammeth F, Ratinger T. The potential of European Union policies to address soil degradation in agriculture. *Land Degrad Dev* 2011;**22**:5–17.
167. Cammeraat LH. Scale dependent thresholds in hydrological and erosion response of a semi-arid catchment in southeast Spain. *Agric Ecosyst Environ* 2004;**104**:317–32.

168. Stavi I, Rozemberg T, Al-ashhab A, Argaman E, Goser E. Failure and collapse of ancient agricultural stone terraces: on-site effects on soil and vegetation. *Water* 2018; **10**:1400.
169. Tarolli P, Preti F, Romano N. Terraced landscapes: from an old best practice to a potential hazard for soil degradation due to land abandonment. *Anthropocene* 2014;**6**:10–25.
170. García-Ruiz JM, Lasanta-Martínez T, Sobrón-García I. Problemas de evolución geomorfológica en campos abandonados: el Valle del Jubera (Sistema Ibérico). *Zubía* 1988; **6**:99–114.
171. Agnoletti M, Conti L, Frezza L, Monti M, Santoro A. Features analysis of dry stone walls of Tuscany (Italy). *Sustainability* 2015;**7**:13887–903.
172. Caneva G, Galotta G, Cancellieri L, Savo V. Tre roots and damages in the Jewish catacombs of Villa Torlonia (roma). *J Cult Herit* 2009;**10**:53–62.
173. Latocha A. Geomorphic connectivity within abandoned small catchments (Stolowe Mts, SW Poland). *Geomorphology* 2014;**212**:4–15.
174. Arnáez J, Lana-Renault N, Ruiz-Flaño P, Pascual N, Lasanta T. Mass soil movement on terraced landscape of the Mediterranean mountain areas: a case study in the Iberian Range, Spain. *Cuadernos de Investigación Geográfica* 2017;**43**(1):83–100.
175. Nadal-Romero E, García-Ruiz JM. Rethinking spatial and temporal variability of erosion in badlands. In: Nadal-Romero E, Martínez-Murillo JM, Khun N, editors. *Badlands dynamics in a context of global change*. Elsevier; 2018. p. 217–53.
176. Pallarés Bou J. Procesos que conducen a la rotura de muros en terrazas de cultivo (Norte de Castellón). *Cuaternario & Geomorfología* 1994;**8**(3–4):23–36.
177. Costra GB, Dal Negro P, Frattini P. Soil slips and debris flows on terraced slopes. *Nat Hazards Earth Syst Sci* 2003;**3**:31–42.
178. Van Dijk AIJM, Bruijnzeel LA, Vertessy RA, Ruijter J. Runoff and sediment generation on bench-terraced hillsides: measurements and up-scaling of a field-based model. *Hydrol Process* 2005;**19**:1667–85.
179. Romero-Martín L, Ruiz-Flaño P, Hernández-Calvento L. Diagnóstico y calidad para la conservación de los espacios agrícolas abancalados. Propuesta metodológica para la cuenca del Guiniguada (Gran Canaria, Islas Canarias). *Geographicalia* 2004;**45**:113–27.
180. Yannelli FA, Tabeni S, Mastrantonio LE, Vezzani N. Assessing degradation of abandoned farmland for conservation of the Monte Desert Biome in Argentina. *Environ Manag* 2014;**53**:231–9.
181. Ries JB, Hirt U. Permanence of soil surface crusts on abandoned farmland in the Central Ebro Basin/Spain. *Catena* 2008;**72**(2):282–96.
182. López-Bermúdez F, Romero-Díaz MA. Piping erosion and badland development in South-east Spain. *Catena Suppl* 1989;**14**:59–73.
183. Lasanta T, García-Ruiz JM, Pérez-Rontomé MC, Sancho C. Runoff and sediment yield in a semi-arid environment: the effect of land management after farmland abandonment. *Catena* 2000;**38**:265–78.
184. Lasanta T, Nadal-Romero E, García-Ruiz JM. Clearing shrubland as a strategy to encourage extensive livestock farming in the Mediterranean mountains. *Cuadernos de Investigación Geográfica* 2019;**45**.
185. Cogliastro A, Gognon D, Daigle S, Bouchard A. Improving harwood afforestation success: an analysis of the effects of soil properties in southwestern Quebec. *Ecol Manag* 2003;**177**:347–59.
186. Benjamin K, Bouchard A, Domon G. Managing abandoned farmland: the need to link biological and sociological aspects. *Environ Manag* 2008;**42**:603–19.
187. Conti G, Fagarazzi L. Sustainable mountain development and the key-issue of abandonment of marginal rural areas. *Planum* 2004;**11**:1–20.

188. Nogués-Bravo D, Simberloff D, Rahbek C, Sanders NJ. Rewilding is the new Pandora's box in conservation. *Curr Biol* 2016;**26**:R83–101.
189. Lasanta T. Active management against shrubland expansion: seeking a balance between conservation and exploitation in the mountains. *Cuadernos de Investigación Geográfica* 2019;**45**. https://doi.org/10.18172/cig.3726.
190. Soulé M, Noss R. Rewilding and biodiversity: complementary goals for continental conservation. *Wild Earth* 1998;**8**:19–28.
191. Rey-Benayas JM, Martins A, Nicolau JM, Schulz JJ. Abandonment of agricultural land: an overview of drivers and consequences. *CAB Rev* 2007;**2**(057).
192. Cramer VA, Hobbs RJ, Standish RJ. What's new about old fields? Land abandonment and ecosystem assembly. *Trends Ecol Evol* 2008;**23**(2):104–12.
193. Ceausu S, Hofmann M, Navarro lH, Carver S, Verburg PH, Pereira HM. Mapping opportunities and challenges for rewilding in Europe. *Conserv Biol* 2015;**29**(4): 1017–27.
194. San Miguel-Ayanz AR, Perea-García-Calvo R, Fernández-Olalla M. Wild ungulates vs extensive livestock. Looking back to face the future. *Options Méditerranéennes A* 2010:92. The contribution of grasslands to the conservation of Mediterranean biodiversity.
195. Moreira F, Queiroz AI, Aronson J. Restoration principles applied to cultural landscapes. *J Nat Conserv* 2006;**14**:217–24.
196. Quetier F, Marty P, Lepart J. Farmer's management strategies and land use in an agropastoral landscape: Roquefort cheese production rules as driver of change. *Agric Syst* 2005;**84**:171–93.
197. Kizos T, Koulouri M. Agricultural landscape dynamics in the Mediterranean: Lesvos (Greece) case study using evidence from the last three centuries. *Environ Sci Policy* 2006;**9**:330–42.
198. Vanslemsbrouck I, Van Huylenbroek G, Van Meensel J. Impact of agriculture on rural tourism: a hedonic pricing approach. *J Agric Econ* 2005;**56**:17–30.
199. Sayadi S, González-Roa MC, Calatrava Requena J. Public preferences for landscape features: the case of agricultural landscapes in mountainous Mediterranean areas. *Land Use Pol* 2009;**26**:334–44.
200. Kinzing AP, Ryan P, Etienne M, Allison H, Elmquist T, Walker B. Resilience and regime shifts: assessing cascading effects. *Ecological Society* 2006;**11**(1):20.
201. Pelorosso R, Della Chiesa S, Tappeiner U, Leone A, Rocchini D. Stability analysis for defining management strategies in abandoned mountain landscapes of the Mediterranean basin. *Landsc Urban Plan* 2011;**103**:335–46.
202. Lasanta T, Nadal-Romero E, Errea MP, Arnáez J. The effects of landscape conservation measures in changing landscape patterns: a case study in Mediterranean mountains. *Land Degrad Dev* 2016;**27**:373–86.
203. Lasanta T, Khorchani M, Pérez-Cabello F, Errea P, Sáenz-Blanco R, Nadal-Romero E. Clearing shrubland and extensive livestock farming: active prevention to control wildfires in the Mediterranean mountains. *J Environ Manag* 2018;**227**:256–66.
204. Lehner B, Doll P, Alcamo J, Henrichs T, Kaspar F. Estimating the impact of global change on flood and drought risks in Europe: a continental, integrated analysis. *Clim Chang* 2006;**75**:273–99.
205. Ortigosa Izquierdo LM. Las repoblaciones forestales como estrategia pública de intervención en regiones degradadas de montaña. In: García-Ruiz JM, editor. *Geoecología de las áreas de montaña*. Logroño: Geoforma Ediciones; 1999. p. 297–311.
206. Pausas JG, Bladé C, Valdecantos A, Seva JP, Fuentes D, Alloza JA, Vilagrosa A, Bautista S, Cortina J, Vallejo R. Pines and oaks in the restoration of Mediterranean landscapes: new perspectives for an old practice. *Plant Ecol* 2004;**171**(1–2):209–20.

207. *S.C.F.C. Informe 2010: Situación de los bosques y del sector forestal en España*. Sociedad Española de Ciencias Forestales; 2011. p. 302.
208. Romero-Díaz A, Belmonte-Serrato F. *Erosión en forestaciones aterrazadas en medio semiárido: Región de Murcia*. Murcia: Universidad de Murcia; 2008. p. 191.
209. Zhetof JHT, Cammeraat ELH, Nadal-Romero E. The enhancing effect of afforestation over secondary succession on soil quality under semiarid climate conditions. *Sci Total Environ* 2019;**652**:1090–101.
210. Nadal-Romero E, Cammeraat E, Pérez-Cardiel E, Lasanta T. How do soil organic carbon stocks change after cropland Mediterranean humid mountain areas? *Sci Total Environ* 2016;**566–567**:741–52.

CHAPTER FOUR

Environments affected by fire

Paulo Pereira[a,*], Juan F. Martínez-Murillo[b,c] and Marcos Francos[d]
[a]Environmental Management Center, Mykolas Romeris Univeristy, Vilnius, Lithuania
[b]Departamento de Geografía, Universidad de Málaga, Malaga, Spain
[c]Instituto de Geomorfología y Suelos (IGSUMA), Universidad de Málaga, Malaga, Spain
[d]Departamento de Ciencias Históricas y Geográficas, Universidad de Tarapacá, Arica, Chile
*Corresponding author: E-mail: pereiraub@gmail.com

Contents

Abstract

Fire is an important element of earth system. It is considered a "global herbivore", a force of nature that shaped the ecosystems and made us "humans" and survive until today. Despite this, nowadays the society understands fire as an evil with extremely negative impacos on the ecosystems. Fire has an important ecological agent, however, can induce soil degradation, especially after high severity fires and recurrent fires. The current sate of our forests, management strategies (e.g. fire suppression, monocultures plantations) and the increasing severity and length of drought spells as consequence of climate change. These unwanted high severity and recurrent fires have detrimental impacts on soil properties and induce a long term degradation (e.g reduction of soil fertility and erosion rates increase). Post-fire restoration is a common practice to reduce

Advances in Chemical Pollution, Environmental Management and Protection, Volume 4
ISSN 2468-9289
https://doi.org/10.1016/bs.apmp.2019.09.001

erosion rates, nevertheless, these should only be applied in areas affected by high severity fires, since normally, in areas burned at low to moderate fire severity, and vegetation recovers fast. In this chapter we will review the impacts of fire on soil properties and the post-fire restoration strategies used.

Keywords: Fire; Environments; Climate change soil degradation; Post-fire treatment.

1. Fire worldwide and environments

Fire is a natural phenomenon that shaped earth biomes as we know today. With the exception of Polar Regions, fire affected with more or less frequency the other biomes. It is considered an ecological force, essential to the existence of several ecosystems (e.g. Savanna, grasslands) and an important element in landscape evolution.[1,2] Without fire, the global distribution of the ecosystems would be tremendously different, and the area of grasslands and savannas area would be much low.[3] Fire is a product of the interaction of three variables, oxygen, fuel and heat and involve a complex set of interactions and feedbacks between biological process in space and time. More recently, the human dimension and their impacts on the landscape became an additional component of this system.[4]

The first evidence of fire was identified 420 million years ago during the Silurian and their frequency increased during the Devonian period. Fires were very common in the Carboniferous and returned in the end of Permian and beginning of Triassic. During these periods fires might be the cause of mass extinctions and affected importantly the earth balance system. During the Pliocene, fire was one of the factors that contributed to the formation of Mediterranean vegetation.[2]

Fire was an element essential to human evolution.[5] We used fire for different purposes: acceleration of biogeochemical cycles, clearing vegetation for cultivation, creating new pastures, hunting, scaring off predators, keeping warm, controlling pests, raw material manipulation (e.g iron, cooper), cooking and burning vegetation to make the landscape habitable. The earliest evidence of the use of fire by humans dates back to 400,000 years.[6] Later, during the twentieth century, because of the abandonment of rural areas, the emergence of an urban society, important interests in rural areas for forest exploitation (e.g. monocultures) and media impact, we started to have a negative vision about fire. Nowadays fires are perceived by the public opinion as an evil element that have negative impacts on the ecosystems.[6–8]

The impact of fire on the ecosystems depends on the severity and regime.[9,10] The highest number of ignitions is observed in Sub-Saharan Africa[11] (Fig. 1), the same is observed in relation to the burned area (Fig. 2) and the occurrence of large wildfires (>500 ha) (Fig. 3). Nevertheless, large fires occurred in other parts of the world, such as Europe,[12] Australia,[13] America[14] and Asia,[15] causing an important environmental, social and economic destruction. It is well known that high severity fires are the ones that have more impacts on soil degradation. Low and medium severity fires do not have serious impacts on the environment, and normally

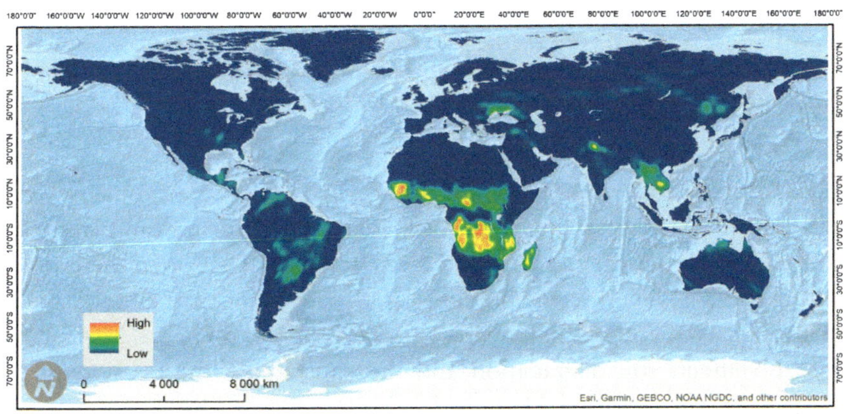

Fig. 1 Density of the fire ignitions in 2016. *Data source (https://glihtdata.gsfc.nasa.gov/files/fire_atlas/, Andela et al.[11]).*

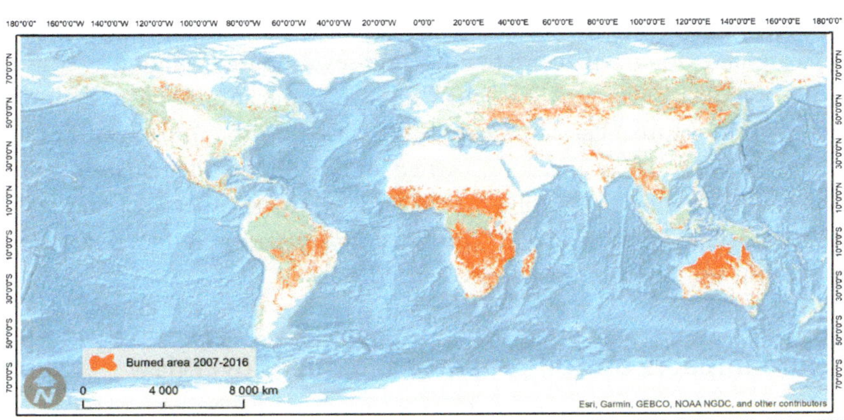

Fig. 2 Burned area between 2007 and 16. *Data source (https://glihtdata.gsfc.nasa.gov/files/fire_atlas/, Andela et al.[11]).*

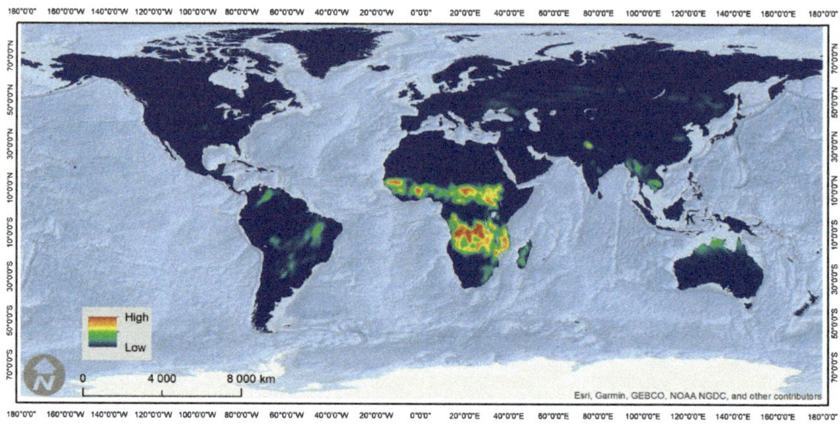

Fig. 3 Density of large wildfires (>500 ha) between 2007 and 16. *Data source (https://glihtdata.gsfc.nasa.gov/files/fire_atlas/, Andela et al.[11]).*

the ecosystem recuperates fast to this disturbance.[16,17] For example, several ecosystems (e.g. Mediterranean) are well adapted to fire and can plants can resprout,[18] while others disperse their seeds when the heat open their cones (serotinity).[19,20] Nevertheless, previous works advocate that some plant or animal species depend on high severity fires for their existence.[21] Fire recurrence also affects soil degradation and the capacity of the ecosystems to recover to fire disturbance. Highest the fire frequency, lowest the capacity of the ecosystem to recuperate and this has been observed in several ecosystems such as the Mediterranean[22] and tropical.[23] It is expected that fire severity and recurrence and the number of megafires will increase with climate change.[24,25] The recent wildfires seasons in Europe, Brazil and Russia are an evidence that fire regimes and severity are changing as consequence of global warming. This shift will have implications on the response of soil to fire disturbance. This chapter is focused on the impacts of fire on soil and the restoration measures carried out to reduce degradation. The work will be focused on soil physical and chemical aspects of the soils. Recently detailed reviews about the fire impacts and post-fire restoration practices on soil microbiology were carried out.[26,27]

1.1 Fire impacts on soil properties: general overview

Soils are an essential component of ecosystems, and can be affected directly (e.g soil heating) and indirectly (e.g ash). With the exception of smoldering fires or log burn, the direct impact of fire is confined to the first centimeters of soil, while the indirect impacts depend more of soil properties (e.g texture) and ash redistribution post-fire. Ecosystem responses after fire depend on pre-fire land use and management, severity, ecosystem affected, topography (e.g., altitude, slope inclination and aspect), soil type, post-fire weather, post-fire management and fire recurrence soils return to pre-fire conditions. Soils covered by flammable industrial plantations, affected by a high severity, located in south faced sloped areas are more vulnerable to degradation and have more difficulties to recover. Soil status in post-fire environments is a key element for ecosystem recovery. Nevertheless, is important to highlight that the high spatial variability nature of fire impacts on soil, as consequence of different types of fuel flammability, connectivity, package, soil type and conditions (e.g. moisture) and topography of the fire affected area, will result in a patchy vegetation recuperation.[26,28–30]

As consequence of the depletion of vegetation, fire can be considered during a certain period (window of disturbance) a geomorphic factor controlling erosional and depositional cycles. Fire influences the evolution and dispersion of plants, biomes and hydrological and erosional cycles. By changing soil infiltration, wettability and runoff generation fire changes sedimentation processes in catchments for some period.[31–34] Nevertheless, some of the fire impacts on soil can be considered permanent and therefore can be considered a formation factor.[35]

1.1.1 Ash layer

Following wildfire or prescribed fire ash commonly cover soil surface. This residue can have a long term impact on the ecosystems.[36] In the immediate period after fire, ash and soil act as a double layer.[29,37] According to Brook and Wittenberg,[38] ash is a unique substance with different mineralogical, textural and structure characteristics. Nevertheless, is very difficult to separate the effects of ash from the effects of soil heating in post-fire environments, especially after the first rainfalls, where they act as one. In post-fire environments, the soil and vegetation recuperation depend on the interaction between the effects of ash and soil in different environmental conditions (e.g. topography, soil type, post-fire weather).

Ash acts like a mulch and protects the soil against erosive agents and crucial for vegetation recuperation. The different colors are an indirect indicator of fire affected the ecosystems. Reddish and black ash are an evidence that the fire had a low to mean severity, while gray and white ashes are an indicator of high fire severity. Ash is an extreme mobile material, especially after high severity wildfires, where it is easily removed by wind. Ash produced at low severity wildfires is heavier and is only transported after intense rainfall periods. After the first rainfalls, ash can be also crusted on soil surface (especially in high severity fires) create an impediment to water infiltration. With the time and after successive precipitation/wind events ash is removed from sloped areas. Nevertheless, if in these areas, soil micro topography is heterogeneous, ash can be deposited in small holes, and function as a nutrient reservoir for soils and plant recover. Normally, ash is transported to lowland areas and reach water reservoirs, changing temporarily water physical, chemical and microbiological properties. With the time and the successive leaching events, the contribution of ash to soil nutrients budget and modification of hydrological processes decrease.[39,40] The time of ash residence on soil is higher in low severity fires compared to high severity ones (Fig. 4).

Ash, mineralogical, physical and chemical composition depends on the type and amount of fuel combusted and temperature of exposition. Normally the major changes in minerology occur at high temperatures (>500–600 °C).[41] Fire temperatures reduce litter height and the rate of increase of mass low is especially relevant in flammable species. The most important changes occur between the temperatures of 250–400 °C. Ash color is reddish at 200–300 °C of exposition, due to the increase iron minerals oxidation, black at 300 °C, as consequence of black carbon formation, gray at temperatures between 350 and 450 °C due to the organic material combustion and white at 450–550 °C as result to the transformation of organic into inorganic carbon. Ash texture also changes with the temperate. From 700 to 900 °C, there is a decrease of particulate size. At temperatures higher than 900 °C, the opposite is observed.[41,42] Also, ash wettability changes with temperature of exposition. Normally ash produced at reduced

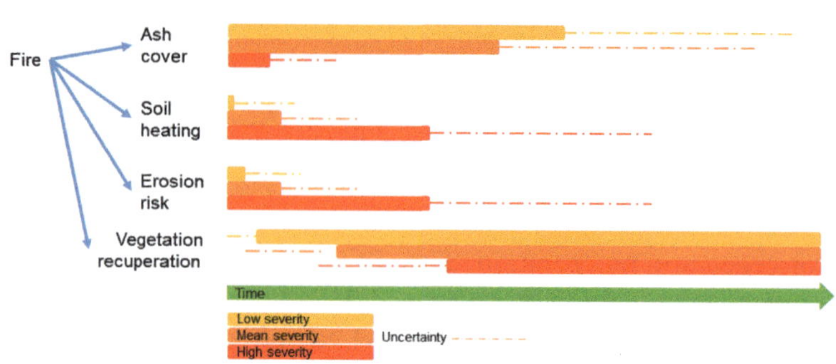

Fig. 4 Conceptual model of fire impacts on ash cover, soil heating, erosion risk and vegetation recuperation.

temperatures (200—400 °C) is more hydrophobic than the ash created at high temperatures (500 °C), and this is attributed to the existence of organic hydrophobic compounds.[43]

The impact of fire temperatures on some nutrients can occur at low temperatures as consequence of the reduced temperatures of volatilization (e.g. carbon and nitrogen start to volatilize at temperatures between 100 and 200 °C). Nevertheless, the majority of the chemical elements only volatilize at high temperatures (>600 °C). Therefore the majority of nutrient losses occur by evacuation during the fire with ash and post-fire, by leaching or erosion.[40,44] Despite the fact that fires increase the losses of some nutrients they became more available as consequence of the mineralization effect of fire. It is well known that fire increases the pH of ash solutions as consequence of the solubility of oxides, hydroxides and carbonates of calcium, magnesium, sodium and potassium. This is especially evident at temperatures higher than 300 °C. This mineralization effect also increase the amount of other ions in solution. Ash extracts are particularly rich in base cations (calcium, magnesium, sodium, potassium). Nevertheless, they contain also heavy metals such as aluminum and manganese that can become available after ash pH decrease.[40,45—47] Ash can be also an important source of polycyclic aromatic hydrocarbons and contribute importantly to the toxicity of soils and water bodies.[48,49]

1.1.2 Impacts on soil physical properties

The impacts of fire on soil physical properties depends on the severity reached. They are more important and prevail more time in high severity fires (Fig. 4). In fires of high severity, the direct impacts (heating) are more important than the indirect impacts (ash), since the ash produced in high severity fires is extremely mobile, contrary to the produced at low and medium fire severity. Nevertheless, is also true that the particle size of the ash created at high severity is finer and can be incorporated easily after the first rainfall periods. However, this impact (depending on the environmental conditions of the burned area and post-fire rainfall) can be short in time (Fig. 4). In areas affected by and low to medium fire severity, the indirect effects of fire may be more relevant. These impacts will decrease with the time and soil properties will return to pre-fire levels.[40,41]

As in the case of ash, soil mass losses increase with the time of exposition as observed om previous works.[50] Soil color changes with the temperature and contact time. Between 25 and 300 °C, soil is black, due the formation of charcoal, while between 300 and 500 °C soil became reddish as consequence of the oxidation of Fe minerals. Normally, the redness ratio increases with the temperature of exposition.[51,52]

At low and medium fire severities, ash can increase soil hydrophobicity.[53–55] Nevertheless, previous works also observed an increase of soil water repellence in these types of fires as consequence of soil heating.[56] In soils affected by high fire severities, temperatures have high impact on mineral soil and is often observed an increase of soil hydrophobicity.[57] However, this depends on the soil type. If the soil is wettable, heating increases water repellency, but if previously to fire the soil is hydrophobic, no changes occur or water repellency can decrease.[58,59] Normally, at temperatures higher than 300–400 °C hydrophobicity is destroyed.[40] Pre-fire vegetation cover also affects soil hydrophobicity. Soils in burned pine plantations are more hydrophobic than in *Quercus coccifera.*[60] Gimeno-Garcia et al.[61] observed that the burned soils under *Rosmarinus officinalis* were more water repellent than under *Q. coccifera.* Other aspect that affects soil water repellency is the moisture content. It is widely known that hydrophobicity decreases with the increasing soil moisture in fire affected soils.[62] Soil water repellency is related with water infiltration, both variables have a high correlation.[63,64] Ash can increase[65,66] or decrease[67] water infiltration depending on their properties. An ash layer enhances soil infiltration by storing water, but when infiltrated can clog soil pores and increase overland flow. Ash rich in carbonates after wetted create a surface crust, and is an impediment for water infiltration.[29] (developed with more detail in Hydrology and soil erosion after fire section).

Changes in soil structure and texture are extremely important since they can affect other soil parameters such as water retention, consistency, infiltration, porosity, heat transfer, and nutrient retention among others.[68] Heating temperatures affect soil texture, especially at high temperatures of exposition as consequence of the fusion of clay particles and modifications of aluminosilicates and oxides and iron hydroxides.[45] For example, Keterings and Bigham[69] found that sand content increased when soil samples were exposed to a temperature higher than 600 °C. Similar results were identified by Roh et al.[70] after heating a soil at 600 °C for 10 min. The authors found an increase in sand size and a decrease in clay particles. Giovannini et al.[71] observed a decrease in clay-size particles after heating a soil at 460 °C for 60 min and Ma et al.[72] found the same trend in soils heated at 700 °C for 20 min. The threshold for the reduction of clay size particles occurs between 350 and 400 °C and this is attributed to the cementation.[73] Nevertheless, other authors identified a reduction of clay content at low temperatures. Badía and Marti[74] observed a decrease of clay size particles after expose a soil at 250 °C for 60 min. This decrease at low temperatures is attributed to the longer time of exposition compared to the previous works. Other authors identified different results. Zhang et al.[76] observed an increase of silt content with increasing temperature of contact. Similar results were identified by Stoof et al.[75] after heating a sample at the temperature of 300 °C.

Fire normally decreases soil aggregate stability as consequence of organic matter combustion,[28] nevertheless the results are not conclusive and there are authors that identified an increase and others a decrease in heated soils, or affected by fire. Low-severity fires do not change or can increase aggregate stability as consequence of the soil water repellency increase.[77] Disaggregation is very likely to occur at high fire temperature/severities. For example Garcia Corona et al.[78] found a high desegregation in a *Leptic Regosol* and a *Leptic Umbrisol*, when exposed to temperatures between 380 and 360 °C. Similar results were identified by Varela et al.[79] and Thomaz and Fachin[80] in *Leptosols*, *Regosols*, *Umbrisols* and *Cambisols*, respectively. On the other hand, Campo et al.[81] did not found significant differences in microaggregates, after heat a *Rendzic Leptosol* at temperatures between 25 and 75 °C. However, in specific conditions at high fire temperature/severity, as consequence aluminum and iron oxides and hydroxides recrystallization, and clay minerals thermal fusion, it is observed an increase of aggregate stability.[82] When a soil is heated to temperatures higher than 500–600 °C, there is an increase of particulates aggregation.[50,80,81,83] Fire residence time and soil moisture have also implications on aggregate stability. Albalasmeh et al.[87] observed that at reduced temperatures and short times of residence, soil aggregates can be disintegrated by the vaporized steam produced by the heating. The soils with high moisture were more vulnerable to aggregate disintegration as consequence of gas-pressure. Since the impacts of temperatures on soil aggregation are complex, in order to have a better understanding of fire effects it is important to consider the analysis of other variables such as aggregate size distribution, soil water repellency, soil organic matter, texture and microbial biomass and fire severity.[82]

Bulk density is strongly related to soil aggregate stability. Some works pointed out that it increases as consequence the decrease in porosity, organic matter or de collapse of mineral aggregates.[85,86] Other works observed that there are no differences between soils affected by low and high fire severity,[87] or even a reduction compared to unburned soils.[39,88] Nevertheless, differences are identified among plant cover and types of soil affected.[89]

Smith et al.[90] found that surface fires in mineral soils did not affect bulk density, while in peat soils there was a significant increase. Munoz-Rojas et al.[91] did not observe significant differences in soils affected by fire in different years. In laboratory experiments, Stoof et al.[75] observed that soil bulk density was significantly high in the sample heated at 300 °C, compared to the control. Wieting et al.[92] found that when soils are heated at 200—250 °C and 450-450 °C, the is an increase of bulk density. Badía and Marti[74] identified an increase of a calcareous and gypsiferous soil bulk density when heated to the temperature of 500 °C. Bulk density was significantly lower in the control sample and at the temperatures of 150 and 250 °C. On the other hand, Zhang et al.[76] did not identify important changes in soil bulk density when they were exposed to temperatures between 20 and 1000 °C.

1.1.3 Impacts on soil chemical properties

One of the most studied soil properties in soil heating experiments or after fire is the pH. In laboratory experiments it is well known that at reduced temperatures of exposition (<300 °C), there is very small changes in pH. At temperatures higher than 300 °C, there is an increase and this is mostly attributed to the formation of oxides, hydroxides and carbonates of calcium and magnesium and organic matter denaturation. The changes of pH after soil heating and fire is very important change because it will rule the bioavailability of nutrients for plant growth and microbiological activity. This is key for landscape recuperation. This increase of pH post-fire is also a consequence of the ash.[41] and it is observed especially in soils affected by high severity.[93,94] Nevertheless changes are also observed in moderate[95,96] and low[97,98] severity fires. In some cases no changes are observed[99,100] and this is attributed to the decrease of soil buffer capacity as consequence of the colloids dehydratation.[40] Soil pH increases with fire severity.[101] pH increase is temporary and decreases after the first rainfalls, despite the fact that some works found that this impact can remain for decades.[102]

As in the case of ash, fire has a mineralization effect on soil organic matter, increasing the amount of ions in solution. In laboratory studies it has been observed that at reduced temperatures of contact (<300 °C), as in pH, the there is no changes in the electrical conductivity. This increase is especially observed in temperatures between 300 and 450 °C, decreasing at high temperatures as consequence of the formation of carbonates, which reduce the amount of ions in solution.[40] The increase of electrical conductivity is also a result of ash incorporation in soil matrix after the first rainfalls.

The increase of electrical conductivity has been reported after low severity fires (e.g prescribed burns),[103–106] while in some cases no changes occurred.[107,108] Increases in soil electrical conductivity are especially observed in medium and high severity fires in the immediate period post-fire as consequence of the intense soil organic matter mineralization. With the time the amount of ions in soil decreases as consequence of erosion, leaching and plant consumption.[109–111]

Fire changes importantly soil organic matter quantity and quality. Heating experiments in laboratory found that organic matter decrease with the increase of the temperatures and time of exposition. Carbon, the most important component of organic matter starts to volatilize at temperatures around 200 °C. The major losses are observed at temperatures between 300 and 450 °C and at temperatures higher than 450 °C, the presence of carbon in soil is reduced.[40] Fire decreases the thickness of the organic layer and the amount of carbon/organic matter is inversely proportional to the severity.[112] Normally after low to moderate severity fires, there is an increase of organic matter or pyrogenic carbon in the mineral soil[112,113] as consequence of the reduced impacts of the temperatures or the incorporation of charred material. Despite this, several works observed a decrease or no changes of soil organic matter[53,114] and this may be attributed to the topography of the burned area (e.g slope inclination), post-fire metereological and reduced vegetation recuperation. The largest losses of organic matter/carbon are observed after high severity fires.[115,116] Organic matter destilation starts at temperatures higher than 100 °C, while the charring processes is initiated at 200 °C. At 300 °C it is observed a decarboxylations of humic and fluvic acids macromolecules and the increase of aromatic compounds. The effects of fire on soil organic matter is dependent on several environmental factors such as severity, soil type, moisture and type of vegetation.[117,118] Fire transforms the existent organic matter chemical structures and create new ones. Jimenez-Morillo et al.[118] found that after a high severity fire, the burned soils the organic compounds were mostly derived from lignin and it was identified the presence of pyrogenic polycyclic aromatic hydrocarbons. The organic matter composition of the unburned soil were mostly composed by lignin proteins and polysaccharides. Merino et al.[119] observed also that fire decreased alkyl and carbohydrates and increased the amount or aromatic compounds. Similar results were identified in other high severity fires.[120] Nevertheless, with the time, the presence of pyrogenic carbon decreases. Alexis et al.[121] found that 11 years after a prescribed fire soil organic matter quality was mostly driven by the inputs of fresh litter very likely due to pyrogenic material degradation, leaching and/or erosion, a process commonly observed in fire affected areas.[122,123]

As consequence of change in carbon cycle, fire affects also the dissolved organic carbon dynamic. Laboratory soil heating experiments found that total dissolved carbon was higher at the temperatures of 150 and 250 °C compared to the unburned sample and 350 °C of exposition. Despite the lower dissolved organic carbon compared with the other temperatures, the aromatic structures were high at 350 °C, which is attributed to the high content of pyrogenic carbon.[125] The works carried out in the field, did not agreed with laboratory experiments. Previous studies observed that low intensity/severity and slash and burn fires can decrease.[126–128] or do not affect[129,130] dissolved organic carbon. The same is observed in relation to soil labile carbon.[131] Some works carried out in wildfire affected areas found a decrease in dissolved organic carbon,[132] while others an increase as observed in a meta analysis carried out by Wang et al.[133] This increase was attributed to the release of carbohydrates due to microbial activity. In any case, as observed in laboratory experiments, the aromacity of the leachates increased in the fire affected soils.[132] The impacts of fire on dissolved organic matter is especially in the quality as consequence of the changes induced by fire on organic matter.[134] Dissolved organic matter after fire is mostly retained in the topsoil[135] and decreases with the soil depth.[55,127]

Soil nitrogen content evolution according to fire temperature/severity is very similar to carbon, as observed in different laboratory experiments. Nitrogen losses increases with temperature and fire severity.[41] Nitrogen volatilize at low temperatures and is easily lost to the atmosphere.[136] Fire induce losses in soil mineral nitrogen content, nevertheless, increases the amount on inorganic nitrogen and the amount of ammonia in solution. This is especially observed at temperatures between 300 and 400 °C, and decreases at temperatures higher than 500 °C.[41] The studies in the field show that low intensity/severity fires did not change[137] or increased available nitrogen forms.[138] In wildfires, normally is observed an increase[139,140] as consequence of the ash deposition and the high soil temperature that stimulates mineralization.[141] In a meta analysis carried out by Wang et al.[128] dissolved total nitrogen was higher in wildfires, compared to prescribed burnings. Nevertheless, not all forms of nitrogen increase. In the immediate period after the fire normally there is a decrease in nitrates in the soil.[139,140] This is followed by an increase in the months after the fire as consequence of nitrification and mineralization process. The increase of nitrates in burned soils compared with unburned ones is especially observed between 3- and 12-months post-fire.[128] The increase of both mineral forms of nitrogen (ammonia and nitrates) is temporal and will decrease with the time as consequence of plant and microbes consumption, adsorption of ammonia in mineral particles with negative charge or can be converted to nitrate due to the increase of nitrifires post-fire.[142]

Soil carbon nitrogen ratio (C:N ratio) decreases with the increasing temperatures. The major reduction is observed at temperatures between 300 and 450 °C. The presence of carbon and nitrogen at temperatures higher than 450 °C is reduced and the losses of these nutrients are positively correlated with mass loss. The decrease of soil C:N ratio after fire is also attributed to the incorporation of ash into the soil matrix.[40,41,117] In the immediate period after the fire, the reduction of C:N ratio is because nitrogen is more phrone to be immobilized than carbon.[116] As consequence of the low temperatures of combustion, immediately after prescribed fires, in some cases there are no changes,[127] however, when high temperatures are observed, there is a significant decrease, especially in the topsoil layer.[114] In wildfires, in the immediate period post-fire, soil C/N ratio is reduced[143] and this can be still observed 20 months after the fire.[144] With the vegetation recuperation, and ash leaching, soil C:N return to pre-fire levels.

Soil cation exchange capacity decreases with the increasing temperatures.[74] Prescribed or low severity fires can reduce[107,147] or not affect[104,138] cation exchange capacity. In moderate and high severity fires, normally is always observed a decrease.[45] This reduction is attributed to the organic matter loss and transformation of clay minerals.[145,146] Nevertheless, ash-bed effect can increase the cation fluxes into soil and increase cation exchange capacity.[148] The changes are especially observed in the first centimeters of the mineral soil and is recovered with time, once vegetation returns fresh organic matter into the soil.[45,107]

Heating temperatures have different impacts on soil the extractable cations. This depends essentially of pH values and carbonate formation. Normally, there is an increase of major cations until 300−450 °C. At higher temperatures it is observed a decrease as consequence of the formation of carbonates, which retain and capture cations in solution. The studies carried out in different types of soils are not conclusive, in some cases there is a high increase of cations, especially major cations. Nevertheless, in others, at reduced temperatures (<300 °C) the concentration of heavy metals (e.g. aluminum, manganese, iron) can be higher in the burned soils compared to the unburned sample. This increase is related to the decrease pH as consequence of colloids dehydratation. Despite different results obtained at low heating temperatures, at high temperatures, especially after 450–500 °C of contact, the number of cations decrease as consequence of the coupled effect of pH and soil carbonate content.[41]

The mineralization effect of fire on litter and soil organic matter increases the availability of major cations (e.g. calcium, magnesium, sodium and potassium), therefore after a fire with low,[107,149] medium and high severity[30,150] is very common to identify an increase of these elements. With the exception of peat fires and pile long burns (the time of residence is high) the impacts of fire are restricted to the topsoil layer, even in fires of high severity. In areas were the time of fire residence is high the impact of fire on soil nutrients is longer.[151,152] The impacts on deeper soil layers is a due to the incorporation of ash and charred material after the first rainfalls.[41]

As consequence of the increase of soil pH, often after fire there is a decrease in the amount of extractable heavy metals. The losses by volatilization are very unlikely since the temperatures of volatilization are very high.[40] Several works observed that the in the immediate period after prescribed fires[153,154] and wildfires[98,155] there is a decrease in the amount of these elements. Nevertheless, the results are not conclusive and previous studies reported also increases of potentially toxic metals such as mercury, cadmium, arsenium, nickel and zinc after a prescribed fire carried out in a legacy mine.[156,157] The increase of heavy metals concentration in soils after wildfires was also reported in other studies.[158,159] This may impose risks to soil and water contamination and affect human health.[160] More works are needed to understand the impact of prescribed and wildfires on heavy metals, since some are essential for plant growth such as iron, zinc, iron, boron, cooper, manganese and nickel.[161–163]

Phosphorous is an element with a very high sensibility to heating. Direct losses occur at temperatures higher than 700 °C. Phosphorous is very phrone to be adsorbed on soil particles in acid and alkaline conditions and is especially soluble between the pH values of 6 and 7. At a pH lower than 6, phosphorous precipitates with iron and aluminum, while at a pH higher than 7 precipitates with calcium.[40,164] This is the reason why in laboratory heating studies phosphorus is more soluble at temperatures between 250 and 450 °C, where the pH values are close to neutrality. At lower temperatures than 250 °C and higher than 450 °C, the presence of phosphorous in burned soils water extracts is residual.[40] Phosphorous availability is strongly linked to ash/soil carbonates content.[165] Fire has a strong impact in phosphorous cycle.[166] The dynamic of phosphorous is complex and the results are not conclusive. Immediately after prescribed fires, extractable phosphorous cannot be affected[138,167] or increase.[112,168,169] A similar situation is observed after wildfires, where some works observed an increase,[96,110] decrease[170] or have no change.[171]

1.2 Importance of fire recurrence

Climate change is increasing fire severity and recurrency inducing an important disturbance on the ecosystems. Fire recurrence the capacity of soils and plants to recover. On the other hand, the increase of fire frequency affects also the global carbon cycle and climate change.[172] Ecosystems such as the Mediterranean are well adapted to fire, and developed several mechanisms (e.g. serotinity or resprout).[10,173,174] Despite this, the change in fire regimen is imposing enormous threats to the resilience of fire phrone ecosystems since is altering profoundly their cycle.[175,176] In general, repeated fires change soil porosity, increase soil bulk density and reduce soil organic matter. This has implications on soil pH, carbon sock, electric conductivity, water and nutrients retention, fine root penetration, nutrients uptake and plant growth, leading to long term decrease of soil productivity and quality.[177]

The disturbance induced by fire on soil properties can be a consequence of prescribed fires used for landscape management or wildfires. In the case of prescribed fires is crucial to find an interval that can be adapted to the different ecosystems. The studies available about the impacts of low prescribed fire interval on soil resources are not consistent and some show that have a detrimental impact, while in others no changes are observed. Differences in the reponses may be related to the ecosystem characteristics (e.g. soil type, vegetation cover, flammability, topography), temperature/severity reached, post-fire weather conditions and vegetation recuperation. Muqaddas et al.[178] observed that a high prescribed fire frequency (2 years) in a native Eucalyptus forest located in Australia, reduced importantly soil moisture, carbon and nitrogen content. Dissolved organic carbon, dissolved organic nitrogen and bioavailable carbon were reduced as well.[179] Also in the US, Wright and Hart[180] identified that a 2-year prescribed fire interval is detrimental for soil carbon and nitrogen. In the same line, Williams et al.[181] found that the application of 5 prescribed fires per decade during a 20 years period reduced importantly soil organic matter, soil organic carbon and increased bulk density. On the other hand, no changes were observed in pH total nitrogen, ammonium, nitrates and extractable potassium. Binkley et al.[182] observed that the management of a loblolly and longleaf pine forest in the Coastal Plain of South Carolina using different prescribed fire intervals (1,2,3 and 4 years) showed that the most frequent fire interval (1 year) reduced the carbon and nitrogen stocks in forest floor. In the Mediterranean environment, Gonzalez-Pelayo et al.[183] found an increase of runoff and soil losses with increasing fire frequency. On the other hand, in the mineral soil, the differences in nitrogen, sulfur, phosphorous, cations and pH were slight. Taylor and Midgley,[184] did not observe changes in soil carbon and phosphorus dynamic and identified a positive feedback in inorganic nitrogen production after an annual prescribed fire in a deciduous hardwood forest (Illinois, the USA). Coates et al.[185] found that in a Longleaf in a forest located in the USA (burned at low severity for 0,4,6 and 8 times between 2004 and 2012), extractable calcium and manganese, was high in the plots burned at higher frequency, while sulfur and iron were high in the unburned plot.

Contrary to the studies carried out in prescribed fires, the increase of wildfire frequency is have detrimental impacts on soil status increasing their degradation. Pellegrini et al.[172] found that soil carbon and nitrogen content in several world ecosystems were decreasing as increasing of wildfire frequency. Similar results were observed by Cheng et al.[186] in grassland and *Acacia* plantations in Taiwan. In Mediterranean environments Mayor et al.[187] observed that soil organic matter and soil fertility decreased with the increasing wildfire severity. Similar results were observed by Keesstra et al.[188] where the increasing frequency of wildfires reduced soil organic matter content. Hosseini et al.[189,190] identified an increased runoff coefficient, soil organic matter, carbon and nitrogen losses after repeated wildfires in *Pinus pinaster* plantations located in Portugal.

2. Hydrology and soil erosion after fire

2.1 Effects of ash layer in soil hydrology

Fire total or partially transform vegetation into ash, which remains covering soil surface for period of time dependent on post-fire weather conditions and topography. The effect of ash in soil hydrology have not been addressed in too many researches as result of the difficulty in examining soil hydrological processes immediately after a wildland fire. The ash layer covering soil surface may act as a layer system, which has an hydrological effect not straightforward and variable rendering in either reducing or increasing overland flow.[29,191,192]

A first influence of ash layer in soil hydrology is its role as storage reservoir for rainfall, preventing overland flow, confirmed in rainfall simulations,[193,194] owing to its high porosity.[195] Consequently, ash-covered soils show less splash detachment as well as overland flow and, thus, erosion and sediment yields.[66,191,196,197] However, other studies have demonstrated that ash can increase overland flow; Bodí et al.[29] highlighted the following factors explaining this process: i) ash depth and type (e.g. composition, particle size, hydrological properties); ii) soil type (e.g. particle size, porosity); and iii) rainfall characteristics (timing, duration and intensity). An ash cover not too much deep besides a longer rainfall event may saturate, pond, and reduce its permeability promoting overland flow by saturation excess mechanisms but also a subsurface flow between the ash and soil.[43,191,195,198,199]

According to Bodí et al.[29] four post-fire scenarios may subsequently occur regarding the ash effect on soil hydrology:

i) Infiltration rate of a two-layer system becomes similar or higher than in baer soil,[195,196,199,200] either reducing overland flow when soil is water repellent and becomes gradually wettable[197,199] and preventing soil from self crusting and compacting.[43,195,198]

ii) Depending on the texture, particle size and pore structure of soil and ash, ash may clog the soil pores, thereby reducing the soil infiltration rates and increasing overland flow.[65,200]

iii) After prolonged or successive rains, although there is a failure in ash structure, this may protects soil from compaction and self-crusting,[191,199] or be entirely removed from a burned site.

2.2 Hydrophobicity and changes in soil infiltration after fire

Hydrophobicity or water repellency is a soil property that describes its water affinity, e.g., the infiltration capacity.[201] Many studies have addressed the controlling factors of hydrophobicity: accumulation of hydrophobic organic compounds from root exudates,[202] soil microorganisms,[203] and soil organic matter decomposition.[204] The hydrophobicity may be registered in soils under a wide range of vegetation types and covers as well as climatic conditions.[204] This soil property displays important hydrological, geomorphological and ecological consequences.[205] Some of them are negative as reduction of soil infiltration rate, increased overland flow and soil erosion, nutrient loss, leaching of fertilizers, and lower seed germination and plant growth[206];

other are positive as enhancing aggregate stability and carbon sequestration.[207]

As other soil properties, hydrophobicity can be altered by fire, either enhanced or destroyed depending on: (i) temperature reached[208]; (ii) time of heating[58]; (iii) type of soil, and quantity and type of fuel.[201] When soil is heating by fire hydrophobicity may be induced owing to partial combustion of organic matter, mainly held in shallow layer of soils,[209] producing volatile organic compounds that condense in the deeper layers, following a negative temperature gradient and coat the mineral soil particles.[208] Consequently, hydrophobicity is one of the soil properties most affected by fire[210] and changes in hydrophobicity may enhance overland flow and erosion in post-fire conditions.[211,212]

2.3 Overland flow and soil erosion in burned hillslopes and catchments

Fire may completely deplete vegetation modifying the hydrological and geomorphic processes acting in hillslopes and catchments: large landscape changes within and downstream of the burned area trigger overland flow and erosion and, subsequent, increments in the frequency of flooding, debris flow, and sedimentation are well documented.[31,213–217]

Post-fire erosion processes are mainly triggered owing to changes in overland flow generation. Fire alters surface hydrology including the following changes: enhancing overland flow generation in response to increments in water repellency[58]; reducing interception of rainfall by forest canopy, understory and litter[217]; either storage of water by ash or increasing overland flow due to ash-related reductions in infiltration[29]; soil surface sealing and crusting due to raindrop impact and, thus, increasing overland flow; incrementing flow velocity as consequence of the reduction in surface roughness; increasing soil saturation due to tree mortality and reducing evapotranspiration; changing snow dynamics[218]; and modifying soil freeze thaw processes, permafrost exposure, and soil creep.

According to Sheridan et al.[217] the magnitude of post-fire effects on soil erosion is controlled by several factors, which may highly vary in space owing to the high spatial variability of fire. These factors are the following:

i) Weather conditions following the fire during the 'window of risk', especially, considering the features of rainfall (total, intensity, form, and timing) and temperature (maximum and timing);

ii) Topography and, especially, gradient, length, and curvature of slopes which control a wide range of erosion processes (splash, interrill, rill, channel, shallow landslide, debris flow, etc.);

iii) Fire severity is strongly related to post-fire erosion as it has been evidenced when compared magnitude orders of erosion rates from low and high severity fire affected areas.[215,219]

Fire imprints the landscape, removing vegetation and either activating or accelerating the geomorphic processes previously mentioned (Fig. 5). After the fire, ash plays a major role as a controller of overland flow generation and soil erosion processes.[220] In many studies it have been observed that not always, immediately after fire, runoff and erosion rates could be negligible due to the ash that covers totally soil surface after a wildfire.[193,196] However, once the ash is removed or crusted into the soil surface due to its mobilization by rainwater drops, overland flow may increase, especially when soils are water repellent. Soil erosion could reach high erosion rates but, in turn, it tends to be reduced owing to the recovery of vegetation cover in time.[199] Extreme and single rain-wash events trigger extraordinary post-fire erosion processes, especially in Mediterranean areas, in autumn, when burned soils are unprotected after summer.[31,221] Nevertheless, this response is not always observed given that soil loss rates are measured either during the first or subsequent years after the wildfire because other factors have to be taken into account. For instance, high rock fragment content on soil surface and within the soil profile, enhancing infiltration and percolation, but also diminishing overland flow velocity.[222–224]

Especially, in Mediterranean conditions, overland flow rate and soil erosion magnitudes are highly variable and erosion rates were generally less than 1 tn ha^{-1} y^{-1}.[225] Scott et al.[226] indicated erosion rates increase

Fig. 5 Examples of geomorphic processes activated after a fire in southern Spain: rill formation in very steep slopes (left); debris flow deposits in skid road (center); and siltation of check-dams (right). *Source: Juan F. Martínez-Murillo.*

one order of magnitude after one fire as well as sedimentation building new landforms very rapidly. Similar results were observed in one study with experimental fires in plots under Mediterranean climatic conditions.[227] It is generally accepted that fire-enhanced erosion rates are maximal immediately after the wildfire (e.g. 35 Mg ha^{-1} during the first postfire year in Fernández et al.[228]) and decrease with time to background levels at the end of the so-called window of disturbance.[31]

3. Post-fire treatments to restore soils

3.1 Detection of target areas to restore burned soils

After a wildfire not all the burned area is equally affected fire. Fire is a common landscape agent in many ecosystems, which usually are adapted to face the vegetation recovery and, even, benefit to certain vegetal species. However, this recovery will not be similar in all the burned area, enhancing soil erosion and making difficult vegetation recovery in post-fire conditions.[229] From the point of view of the post-fire management, the definition of areas with more difficulties for recovery is a key issue. The actions to enhance the vegetation recovery and, thus, to protect soils, may be focused on those priority areas to minimize soil degradation.

Sometimes, when these priority areas are not highlighted and all the burned area is treated similarly soil degradation and erosion processes can be enhanced owing to the post-fire land management. For instance, in one wildfire occurred in South of Spain, affecting a very mountainous area with very steep slopes and shallow and stoniness soils, overland flow promoted rill and gully erosion taking advantage of the new skid trails as well as the salvage logging carried out during the post-fire management (Fig. 6).[230]

Thus, post-fire management is normally focused on minimizing post-fire erosion effects (mitigation) and shortening ecosystem recovery times (rehabilitation).[231] Both post-fire mitigation and rehabilitation treatments can be costly and cumbersome due to the extension of the area affected. Besides, ecosystems affected by fires in summertime (e.g. the Mediterranean ecosystem) are generally followed by torrential rainfalls, causing water erosion and soil degradation. Because of this, post-fire treatments must be performed within weeks after the fire. In this context, it is critical to target the efforts on high priority locations to reduce erosion risks in a cost effectively approach.[231]

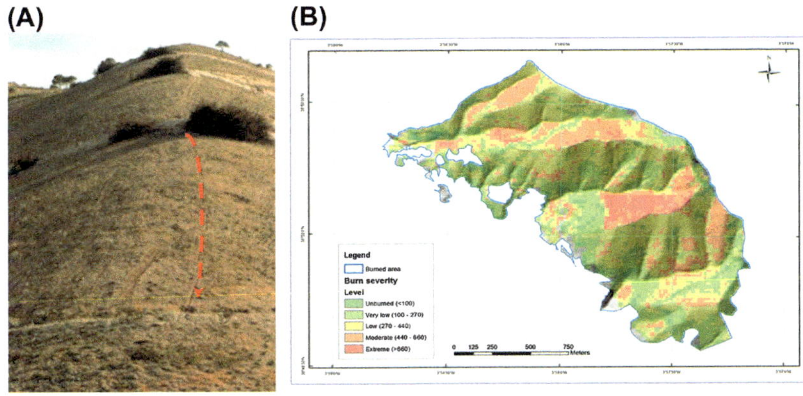

Fig. 6 (A) Soil erosion and formation of rills owing to salvage post-fire management. (B) Example of burn severity map from a burned are in southern Spain. *Source: Juan F. Martínez-Murillo.*

To do this, both short and long-term post-fire effects on vegetation and soil can be estimated in terms of "burn severity".[232–234] The spatial variability of the level of damage and its distribution within the burned area (burn severity map) (Fig. 6) is crucial to quantify the impact of fires on landscapes,[232] to select and prioritize treatments applied on site,[235] to plan and monitor restoration and recovery activities and, finally, to provide baseline information for future monitoring.[236]

Different methods of burn severity estimation have been applied using post-fire field evaluation of soil and vegetation conditions.[237,238] De Santis et al.[234] pointed out field surveys are costly and time consuming and do not provide a good spatial coverage, remote sensing imagery has been proposed as a sound alternative. For instance, Interagency Burned Area Emergency Rehabilitation (BAER) teams use a semi-automatic estimation of burn severity from satellite images on a nationwide scale, applying techniques developed by the Remote Sensing Applications Center - RSAC of the USDA Forest Service. The field assessment method used by BAER is based on the widely used Composite Burn Index (CBI), which takes continuous values ranging from 0 (unburned) to 3 (completely burned). However, in most studies dealing with burn severity estimation from satellite imagery, damage levels are normally grouped into only four severity classes: unburned, low, moderate and severe.[232,239–244] In order to reduce this limitation in accuracy and precision, other methods were proposed using of Radiative Transfer Models (RTMs) to simulate the continuum interval of burn severity levels measured in CBI.[233,234]

3.2 Effects of amendment application on soil properties, vegetation recovery and soil erosion

Direct effects of fires due to soil heating, such as breakdown of aggregates and increased soil water repellency, are generally considered to be key factors in the strong and sometimes extreme hydrological and erosion responses of recently burnt areas.[58,79,245] Fire-enhanced generation of overland flow and the associated export of sediments, organic matter, nutrients and pollutants not only have negative consequences for on-site land use sustainability, but also can endanger downstream aquatic and flood-zone habitats and associated human infrastructures.[31,245,246] Erosion rates may be very high in the following months after fire until vegetation recovery starts decreasing to levels at the end of the window of disturbance. However, the intensity and extent of this period, which depends on fire severity and post-fire climate conditions, are still highly uncertain and difficult to quantify.[247]

Different treatments have been identified to effectively reduce post-fire soil erosion.[248,249] Among all of them, the most widely accepted measure is mulching, i.e., the application of a cover of organic compounds on the soil surface to modify energy and water fluxes and to protect the soil from direct raindrop impact.[250] Mulching has been found to successfully control post-fire runoff and soil erosion in many field trials.[228,246,251] A mulch cover of 60% is widely considered the minimum threshold for a significant reduction in soil loss.[196,249,252,253] In the case of straw mulch, this threshold cover is typically achieved by applying 2 Mg of straw per ha.[74,228,251,255]

Although burned areas are commonly mulched with straw, this may have various disadvantages: high cost, potential introduction of non-native plants, and susceptibility to wind-scattering.[250] Some years ago new mulch types were designed using alternative materials derived from forest residues, using fibers of different shapes and sizes.[256,257] Similarly, in laboratory experiments, 6-cm long wood strands applied at rates of 4–8 Mg ha^{-1} were found to be highly effective, reducing erosion rates by 80%.[258–260] In field trials, mulching with 10- to 15-cm long chopped eucalyptus bark fibers markedly reduced post-fire erosion during the first year after the fire.[246]

Recently, one more measure to control post-fire erosion has been developed with the application of polyacrylamides (PAMs).[249] PAMs refer to a family of flocculant agents, comprising a broad class of chemical compounds with different chain lengths, charge types and charge densities. Different PAM formulations have been developed to ensure effective binding with clay particles through direct ionic attractions or cation bridges.[261,262] The application of PAMs constitutes a remarkable soil- and water-management technique, due to their extremely low cost, safety, and their capacity to influence physicochemical processes.[263] During the last two decades, the use of PAMs has proven effective for erosion control in furrow irrigation in intensive agriculture.[263,264]

4. Conclusions

Fire is a natural element of the ecosystems, however as consequence of complex-socio economic processes and climate change, the frequency and severity of fires is increasing. Low to moderate severity fires can have beneficial or no detrimental effects on soils degradation. The most serious impacts are observed after high severity fires where the high temperatures combust litter layer and soil organic matter. This can lead to a long-term degradation depending on the environmental characteristics of the burned area and post-fire weather conditions. These aspects are crucial to vegetation recuperation, key to the restoration of soil ecosystem services. The increase of fire recurrence, imposes an additional disturbance to soil. If the interval between fires is reduced, the capacity of soil recuperation to fire impact is reduced. This is the beginning of a degradation process. Post-fire interventions should be considered carefully and applied in areas affected by high severity fires and it is essential that these interventions are not carried out in the immediate period after the fire. In the areas affected by low and moderate fire severities vegetation can recover easily and therefore restoration practices are not needed. There are several post-fire restauration measures, normally applied to reduce erosion rates in hill slopes. Among them, mulch is one of the most popular and most effective.

Acknowledgments

This work was supported by the Croatian Science Foundation through the project "Influence of Summer Fire on Soil and Water Quality" (U IP-2018-01-1645).

References

1. Naveh Z. The evolutionary significance of fire in the Mediterranean region. *Vegetatio* 1975;**29**:199–208.
2. Pereira P, Rein G, Martin D. Editorial: past and present post-fire environments. *Sci Total Environ* 2016;**573**:1275–7.
3. Bond WJ, Woodward FI, Midgley GF. The global distribution of ecosystems with a world without fire. *New Phytol* 2004;**165**:525–38.

4. Harris RMB, Remenyi TA, Williamson GJ, Bindoff NL, Bowman DMJS. Climate-vegetation-fire and feedbacks: trivial detail or major barrier to projecting the future of the earth system. *Wires Clim Change* 2016;**7**:910–31.
5. Pausas JG, Keeley JE. A burning history: the role of fire in the history of life. *Bioscience* 2009;**59**:593–601.
6. Santín C, Doerr SH. Fire effects on soils: the human dimension. *Phil. Trans. R. Soc. B* 2016;**371**:20150171.
7. Regos A, Dominguez J, Gil-Tena A, Brotons L, Ninyerola M, Pons X. Rural abandoned landscapes and bird assemblages: winners and losers in the rewilding of a marginal mountain area (NW Spain). *Reg Environ Chang* 2016;**16**:199–211.
8. Velez ALK, Diaz JM, Wall TU. Public information seeking, place based risk messaging and wildfire preparedness in southern California. *Int J Wildland Fire* 2017;**26**:469–77.
9. Basset M, Leónard SWJ, Chia EK, Clarke MF, Bennett AF. Interacting effects of fire severity, time since fire and topography on begetation structure after wildfire. *For Ecol Manag* 2017;**369**:26–34.
10. Fernandez-Garcia V, Fule PZ, Marcos E, Calvo L. The role of fire severity on the regeneration of Mediterranean serotineous pines under different environment conditions. *For Ecol Manag* 2019;**444**:59–68.
11. Andela N, Morton DC, Giglio L, Paugam R, Chen Y, Hanson S, van der Werf GR, Randerson JT. The Global Fire Atlas of individual fire size, duration, speed, and direction. *Earth Syst Sci Data* 2019;**11**:529–52.
12. Diakakis M, Nikolopoulos E, Mavroulis S, Vassilakis E, Korakaki E. Observational evidence on the effects of megafires on the frequency of hydrogeomorphic hazards. The case of the Peoloponnese fires of 2007 in Greece. *Sci Total Environ* 2017;**592**:262–72.
13. French BJ, Prior LD, Williamson GJ, Bowman DMJS. Cause and effects of a megafire in sedge-heathland in Tasmanian temperate wilderness. *Aust J Bot* 2016;**64**:513–25.
14. De la Barrera F, Barraza F, Favier P, Ruiz V, Quense J. Megafires in Chile 2017: monitoring multiscale environmental impacts of burned ecosystems. *Sci Total Environ* 2019; **637–638**:1526–36.
15. Zhu C, Kobayashi H, Kanaya Y, Saito M. Size-dependent validation of MODIS MCD64A1 burned area over six vegetation types in boreal Eurasia. Large underestimation in croplands. *Sci Rep* 2017;**7**:4181.
16. Gonzalez de la Vega S, De las Heras J. Resilience of Mediterranean terrestrial ecosystems and fire severity in semiarid areas: responses of Aleppo pine forests in the short, mid and long term. *Sci Total Environ* 2016;**573**:1171–7.
17. Viana-Soto A, Aguado I, Martinez S. Assessment of post-fire severity vegetation recovery using fire severity and geographical data in the Mediterranean region. *Environments* 2017;**90**.
18. Pausas JG, Keeley JE. Epicormic resprouting in fire-prone ecosystems. *Trends Plant Sci* 2017;**22**:1008–15.
19. Tapias R, Climent J, Pardos JA, Gil L. Life histories of Mediterranean pines. *Plant Ecol* 2004;**171**:53–68.
20. Canelles Q, Aquilue N, Duane A, Brotons L. From stand to landscape: modelling post-fire regeneration and species growth. *Ecol Model* 2019;**404**:103–11.
21. Hutto RL, Keane RE, Sherriff RL, Rota CT, Eby LA, Saab VA. Toward more ecologically informed view of severe forest fires. *Ecosphere* 2016;**7**:e01255.
22. Tessler N, Wittenberg L, Greenbaum N. Vegetation cover and species richness after recurrent fires in the Eastern Mediterranean ecosystem of Mount Carmel, Israel. *Sci Total Environ* 2016;**572**:1395–402.
23. Flores BM, Fagoaga R, Nelson BW, Holmgren M. Repeated fires trap Amazonian blackwater floodplains in open vegetation state. *J Appl Ecol* 2016;**53**:1597–603.

24. Seidl R, Thom D, Kautz M, Martin-Benito D, Peltoniemi M, Vacchiano G, Wild J, Ascoli D, Petr M, Honkaniemi J, Lexer MJ, Trotsiuk V, Mairota P, Svoboda M, Fabrika M, Nagel TA, Reyer CPO. Forest disturbances under climate change. *Nat Clim Chang* 2017;**7**:395–402.
25. Schoennagel T, Balch JK, Brenkert-Smith H, Dennison PE, Harvey BJ, Krawchuk MA, Mietkiewicz N, Morgan P, Moritz MA, Rasker R, Turner MG, Whitlook C. Adapt to more wildfire. *Proc Natl Acad Sci USA* 2017;**144**:4582–90.
26. Pereira P, Francos M, Brevik EC, Úbeda X, Bogunovic I. Post-fire soil management. *Curr Opin Environ Sci Health* 2018;**5**:26–32.
27. Munoz-Rojas M, Barcenas-Moreno. Microbiology. In: Pereira P, Mataix-Solera J, Úbeda X, Rein G, Cerdà A, editors. *Fire effects on soil properties*. CSIRO Publishing (Australia); 2019. p. 15–38.
28. Badía D, González-Pérez JA, Aznar JA, Arjona-Garcia B, Martí-Dalamau C. Changes in soil water repellency aggregation and organic matter of a mollic horizon burned in laboratory: soil depth affected by fire. *Geoderma* 2014;**213**:400–7.
29. Bodí M, Martin DA, Balfour VN, Santín C, Doerr SH, Pereira P, Cerdà A, Mataix-Solera J. Wildland fire ash: production, composition and eco-hydro-geomorphic effects. *Earth Sci Rev* 2014;**130**:103–27.
30. Caon L, Vallejo VR, Coen RJ, Geissen V. Effects of wildfire on soil nutrients in Mediterranean ecosystems. *Earth Sci Rev* 2014;**139**:47–58.
31. Shakesby RA, Doerr SH. Wildfire as a hydrological and geomorphological agent. *Earth Sci Rev* 2006;**74**:269–307.
32. Martinez-Murillo J, Neris J, Hyde K, Keiser. Advances towards an integrated assessment of fire effects on soil, vegetation and geomorphological processes. *Land Degrad Dev* 2016;**27**:1314–8.
33. Langhans C, Nyman P, Noske PJ, Van der Sant RE, Lane PNJ, Sheridan GJ. Post-fire hillslope debris flows: evidence of a distinct erosion process. *Geomorphology* 2017;**295**: 55–75.
34. Brogan DJ, MacDonald LH, Nelson PA, Morgan JA. Geomorphic context and sensitivity in channels fo fire and floods in mountain catchments. *Geomorphology* 2019;**337**: 53–68.
35. Certini G. Fire as a soil forming factor. *Ambio* 2014;**43**:191–5.
36. Pereira P, Jordán A, Cerdà A, Martin D. Editorial: the role of ash in fire-affected ecosystems. *Catena* 2015;**135**:337–9.
37. Brook A, Wittenberg A. Ash-soil interface: mineralogical composition and physical structure. *Sci Total Environ* 2019;**572**:1403–13.
38. Brook A, Wittenberg L, Kopel D, Polinova M, Roberts D, Ichoku C, Shtober-Zisu N. Structural heterogeneity of vegetation fire ash. *Land Degrad Dev* 2018;**29**:2208–21.
39. Goforth BR, Graham RC, Hubbert KR, Zanner CW, Minnich RA. Spatial distribution and properties of ash and thermally altered soils after high-severity forest fire, southern California. *Int J Wildland Fire* 2005;**14**:343–54.
40. Pereira P, Brevik E, Bogunovic I, Estebaranz-Sánchez F. Ash and soils. A close relationship in fire-affected areas. In: Pereira P, Mataix-Solera J, Úbeda X, Rein G, Cerdà A, editors. *Fire effects on soil properties*. CSIRO Publishing (Australia); 2019a. p. 15–38.
41. Pereira P, Úbeda X, Francos M, Estebaranz-Sánchez F. Ash and soils. A close relationship in fire-affected areas. In: Pereira P, Mataix-Solera J, Úbeda X, Rein G, Cerdà A, editors. *Fire effects on soil properties*. CSIRO Publishing (Australia); 2019b. p. 15–38.
42. Gray DM, Dighton J. Mineralization of forest litter nutrients by litter and combustión. *Soil Biol Biochem* 2006;**38**:1469–77.

43. Bodí MB, Mataix-Solera J, Doerr SH, Cerdà A. The wettability of ash from burned vegetation and its relationship to Mediterranean plant species type, burn severity and total organic carbon content. *Geoderma* 2011;**160**:599–607.
44. Knicker H. How does fire affect the nature and stability of soil organic nitrogen and carbón? A review. *Biogeochemisty* 2007;**85**:91–118.
45. Ulery AL, Graham RC. Forest fire effects on soil color and texture. *Soil Sci Soc Am J* 1993;**57**:135–40.
46. Henig-Sever N, Poliakov D, Broza M. A novel method for estimation of wildfire severity base don ash pH and microantropod. *Pedobiologia* 2011;**45**:98–106.
47. Úbeda X, Pereira P, Outeiro L, Martin DA. Effects of fire temperature on the physical and chemical characteristics of the ash from two plots of cork oak (Quercus suber). *Land Degrad Dev* 2009;**20**:589–608.
48. Campos I, Abrantes N, Vidal T, Bastos AC, Goncalves F, Keiser JJ. Assessment of the toxicity of ash-loaded runoff from a recently burnt eucalypt plantation. *Eur J For Res* 2012;**131**:1889–903.
49. Campos I, Abrantes N, Pereira P, Micaelo AC, Vale C, Keizer JJ. Forest fires as potential triggers for production and mobilization of polycyclic aromatic hydrocarbons to terrestrial ecosystem. *Land Degrad Dev* 2019. https://doi.org/10.1002/ldr.3427.
50. Jimenez-Pinilla P, Mataix-Solera J, Arcenegui V, Delgado R, Martin-Garcia JM, Lozano E, Martinez-Zavala L, Jordán A. Advances in the knowledge of how heating can affect aggregate stability in Mediterranean soils: a XRD and SEM-EDX approach. *Catena* 2016a;**147**:315–24.
51. Wondafrash TT, Mariscal Sancho I, Gomez Miguel I, Espejo Serrno R. Relationship between soil color and temperature in a surface horizon of Mediterranean soils: a laboratory study. *Soil Sci* 2005;**170**:495–503.
52. Terefe T, Mariscal-Sancho I, Peregrina F, Espejo R. Influence of heating on various properties of six Mediterranean soils. *Geoderma* 2008;**143**:273–80.
53. Pereira P, Úbeda X, Mataix-Solera J, Oliva M, Novara A. Short-term changes in soil Munsell colour value, organic matter content and soil water repellency after a spring grassland fire in Lithuania. *Solid Earth* 2014;**5**:209–25.
54. Jimenez-Pinilla P, Lozano E, Mataix-Solera J, Arcenegui V, Jordán A, Zavala LM. Temporal changes in soil wáter repellency after forest fire in a Mediterranean calcareous soil: influence of ash and different vegetation type. *Sci Total Environ* 2016;**572**: 1252–60.
55. Girona-Garcia A, Ortiz-Perpina O, Badía-Villas D, Marti-Dalmau C. Effects of prescribed burning on soil organic C, aggregate stability and wáter repellency in a subalpine shrubland: variations among sieve fractions and depths. *Catena* 2018;**166**: 68–77.
56. Cawson JG, Nyman P, Smith HG, Lane PNJ, Sheridan GJ. How soil temperatures during prescribed burning affect soil wáter repellency, infiltration and erosion. *Geoderma* 2016;**278**:12–22.
57. MacDonald LH, Huffman EL. Post-fire soil water repellency. *Soil Sci Soc Am J* 2004; **68**:1729–34.
58. Doerr SH, Shakesby RA, Walsh RPD. Soil water repellency: its causes, characteristics and hydrogeomorphological significance. *Earth Sci Rev* 2000;**51**:33–65.
59. Badía D, Lopez-Garcia S, Marti C, Ortiz-Perpina O, Girona-Garcia A, Casanova-Gascon J. Burn effects on soil properties associated to heat transfer under contrasting moisture content. *Sci Total Environ* 2017;**601–602**:1119–28.
60. Mataix-Solera J, Arcenegui V, Guerrero C, Mayoral AM, Morales J, González J, García-Orenes F, Gómez I. Water repellency under different plant species in a calcareous forest soil in a semiarid Mediterranean environment. *Hydrol Process* 2007;**21**:2300–9.

61. Gimeno-Garcia E, pascual JA, Llovet J. Water repellency and moisture content spatial variations under *Rosmarinus officinalis* and *Quercus coccifera* in Mediterranean burned soil. *Catena* 2011;**85**:48–87.
62. Plaza-Alvarez PA, Lucas-Borja ME, Sagra J, Moya D, Alfaro-Sanchez R, Gonzalez-Romero J, De las Heras J. Changes in soil water repellency after prescribed burnings in three different Mediterranean forest ecosystems. *Sci Total Environ* 2018;**644**:247–55.
63. Lewis SS, Wu JQ, Robichaud PR. Assessment of burn severity and comparing soil wáter repellency, Hayman Fire, Colorado. *Hydrol Process* 2006;**20**:1–16.
64. Robichaud PR, Wagenbrenner JW, Pierson FB, Spaeth KE, Ashmun LE, Moffet CA. Infiltration and interrill erosion rates after wildfire in western Montana, USA. *Catena* 2016;**142**:77–88.
65. Woods SW, Balfour VN. The effects of soil texture and ash thickness on the postfire hydrological response from ash-covered soils. *J Hydrol* 2010;**393**:274–86.
66. León J, Bodí MB, Cerdà A, Badía D. The contrasted response of ash to wetting. The effects of ash type, thickness and rainfall events. *Geoderma* 2013;**209–210**:143–52.
67. Onda Y, Dietrich WE, Booker F. Evolution of overland flow after a severe forest fire, Point Reyes, California. *Catena* 2008;**72**:13–20.
68. Badía D, Marti C. Texture, minerology and structure. In: Pereira P, Mataix-Solera J, Úbeda X, Rein G, Cerdà A, editors. *Fire effects on soil properties*. CSIRO Publishing (Australia); 2019. p. 69–80.
69. Ketterings QM, Bigham JM. Soil color as an indicator of slash-and-burn fire severity and soil fertility in Sumatra, Indonesia. *Soil Sci Soc Am J* 2000;**64**:1826–33.
70. Roh Y, Edwards NT, Lee SY, Stiles CA, Armes S, Foss JE. Thermal-treated soil for Mercury removal: soil and phytotoxicity tests. *J Environ Qual* 2000;**29**:415–24.
71. Giovannini G, Lucchesi S, Giachetti M. Effect of heating on some physical and chemical parameters related to soil aggregation and erodibility. *Soil Sci* 1998;**146**:255–61.
72. Ma F, Zhang Q, Xu D, Hou D, Li F, Gu Q. Mercury removal from contaminated soil by thermal treatment with $FeCl_3$ at reduced temperature. *Chemosphere* 2012; **117**:388–93.
73. O'Brian PL, DeSutter T, Casey FXM, Khan E, Wick F. Thermal remediation alters soil properties – a review. *J Environ Manag* 2018;**206**:826–35.
74. Badía D, Martí C. Effect of simulated fire on organic matter and selected microbiological properties of two contrasting soils. *Arid Land Res Manag* 2003;**17**:55–69.
75. Stoof CR, Wesseling JG, Ritsema CJ. Effects of fire and ash on soil wáter retention. *Geoderma* 2010;**159**:276–85.
76. Zhang F, Kong R, Peng J. Effects of heating on compositional, structural, and physicochemical properties of loess. *Appl Clay Sci* 2018;**152**:259–66.
77. Fox DM, Darboux F, Carrega P. Effects of fire-induced wáter repellency on soil aggregate stability, splash erosion, and saturated hydraulic conductivity for different size fractions. *Hydrol Process* 2007;**21**:2377–84.
78. Garcia-Corona R, Benito E, Blas E, Varela ME. Effects of heating on some soil physical properties related to its hydrological behaviour in two north-western Spanish soils. *Int J Wildland Fire* 2004;**13**:195–9.
79. Varela ME, Benito E, Blas E. Impacts of wildfire and laboratory heating on aggregate stability in Pine forests in Galicia: the role of lithology, soil organic matter content and water repellency. *Catena* 2010;**83**:127–34.
80. Thomaz EL, Fachin PA. Effects of heating on soil physical properties by using realistic peak temperatura gradients. *Geoderma* 2014;**230–231**:243–9.
81. Campo J, Gimeno-Garcia E, Andreu V, Gonzalez-Pelayo O, Rubio JL. Cementing agents involved in macro and microaggregation of a Mediterranean shrubland soil under laboratory heating. *Catena* 2014;**113**:165–76.

82. Mataix-Solera J, Cerdà A, Arcenegui V, Jordán A, Zavala LM. Fire effects on soil aggregation: a review. *Earth Sci Rev* 2011;**109**:44–60.
83. Thomaz EL. Fire changes the larger aggregate stability classes in slash-and-burn agricultural systems. *Soil Res* 2017;**165**:210–7.
84. Deleted in review.
85. Certini G. Effects of fire on properties of forest soils: a review. *Oecologia* 2005;**143**:1–10.
86. Hubbert KR, Preisler HK, Wohlgemuth PM, Graham RC, Narog MG. Prescribed burning effects on soil physical properties and soil water repellency in a steep chaparral watershed, southern California, USA. *Geoderma* 2006;**130**:284–98.
87. Jordán A, Zavala LM, Mataix-Solera J, Nava AL, Alanis N. Effect of fire severity on wáter repellency and aggregate stability on Mexican volcanic soils. *Catena* 2011;**84**: 136–47.
88. Shaoqing C, Shaolin P, Baoming C, Danting C, Juhua C. Effects of fire disturbance on the soil physical and chemical properties and vegetation of *Pinus massoniana* forest in south subtropical área. *Acta Ecol Sin* 2010;**30**:184–9.
89. Zavala L, Gonzalez FA, Jordán A. Fire-induced soil wáter repellency under different vegetation types along the Atlantic dune coast-line in SW Spain. *Catena* 2009a;**79**: 153–62.
90. Smith SM, Newman S, Garrett PB, Leeds JA. Differential effects of Surface and peat fire on soil constituents in a degraded wetland of the Northern Florida Everglades. *J Environ Qual* 2001;**30**:1998–2005.
91. Munoz-Rojas M, Erickson TE, Martini D, Dixon KW, Merritt DJ. Soil physicochemical and microbiological indicators of short, médium and long term post-fire recovery in semi-arid ecosystems. *Ecol Ind* 2016;**63**:14–22.
92. Wieting C, Ebel BA, Singha K. Quantifying the effects of wildfire on changes in soil properties by Surface burning of soils from Boulder Creek critical zone. *J Hydrol: Reg Stud* 2017;**13**:43–57.
93. Knelman JE, Graham EB, Trahan NA, Schmidt SK, Nemergut DR. Fire severity shapes plant colonization effects on bacterial community structure, microbial biomass, and soil enzyme activity in secondary succession of a burned forest. *Soil Biol Biochem* 2015;**90**:161–8.
94. Ulery AL, Graham RC, Goforth BR, Hubbert KR. Fire effects on cation exchange capacity of California forest and woodland soils. *Geoderma* 2017;**286**:125–30.
95. Brais S, David P, Ouimet R. Impacts of wild fire severity and salvage harvesting on the nutrient balance af Jack pine and black spruce boreal stands. *For Ecol Manag* 2000;**137**: 231–43.
96. Heydari M, Rostamy A, Najafi F, Dey DC. Effect of fire severity on physical and biochemical soil properties in Zagros oak (*Quercus brantii* Lindl) forests in Iran. *J Res* 2017;**28**:95–104.
97. Akburak S, Son Y, Makineci E, Cakir M. Impacts of low-intensity prescribed fire on microbial and chemical soil properties in a *Quercus frainetto* forest. *J Res* 2018;**29**: 687–96.
98. Pereira P, Cerdà A, Martin D, Úbeda X, Depellegrin D, Novara A, Martinez-Murillo JF, Brevik EC, Menshov O, Comino J, Miesel J. Short-term low severity spring grassland fire impacts on soil extractable elements and soil ratios in Lithuania. *Sci Total Environ* 2017;**578**:469–75.
99. Rhoades C, Barnes T, Washburn B. Prescribed fire and herbicide effect on soil processes during barrens restoration. *Restor Ecol* 2002;**10**:656–64.
100. Savadogo P, Sawadogo L, Tiveau D. Effects of grazing intensity and prescribed fire on soil physical and hydrological properties and pasture yield in the savanna woodlands of Burkina Faso. *Agric Ecosyst Environ* 2007;**118**:80–92.

101. Merino A, Fonturbel MT, Fernandez C, Chavez-Vergara B, Garcia-Oliva F, Vega JA. Inferring changes in soil burn severity levels in a temperate climate. *Sci Total Environ* 2018a;**627**:622–32.
102. Khanna PK, Raison RJ. Effect of fire intensity on solution chemistry of surface soil under a Eucalyptus pauciflora forest. *Aust J Soil Res* 1986;**24**:423–34.
103. Guinto DF, Xu ZH, House APN, Saffigna PG. Soil chemical properties and Forest floor nutrients under repeated prescribed-burning in eucalypt forest of South-East Queensland, Australia. *N Z J For Sci* 2001;**31**:170–87.
104. Arocena JM, Opio C. Prescribed fire-induced changes in properties of sub-boreal forest soils. *Geoderma* 2003;**113**:1–16.
105. Kara O, Bolat I. Short-term of wildfire on microbial biomass and abundance in black pine plantation soils in Turkey. *Ecol Ind* 2009;**9**:1151–5.
106. Granged AJP, Zavala LM, Jordán A, Barcenas-Moreno G. Post-fire evolution of soil properties and vegetation cover in a Mediterranean heathland after a experimental burning: a 3 year study. *Geoderma* 2011;**164**:85–94.
107. Fonseca F, Figueiredo T, Nogueira C, Queiros A. Effect of prescribed fire on soil properties in a Mediterranean mountain área. *Geoderma* 2017;**307**:172–80.
108. Naidu CV, Srivasuki KP. Effect of forest fire on soil characteristics in different areas of Seshachalam hills. *Ann For Sci* 1994;**2**:166–73.
109. Kutiel P, Naveh Z. The effect of fire on nutrients in a pine forest soil. *Plant Soil* 1987; **104**:269–74.
110. Alauzis MV, Mazzarino MJ, Raffaele E, Roselli L. Wildfires in NW Patagonia: long-term effects on Nothofagus forest soil. *For Ecol Manag* 2004;**192**:131–42.
111. Pardini G, Gispert M, Dunjo G. Relative influence of wildfire on soil properties and erosion processes in different Mediterranean environments in NE Spain. *Sci Total Environ* 2004;**328**:237–46.
112. Merino A, Jimenez E, Fernandez C, Fonturbel MT, Campo J, Vega JA. Soil organic matter and phosphorous dynamics after low intensity prescribed burning in forest and shrubland. *J Environ Manag* 2019;**237**:214–25.
113. De Marco A, Gentile AE, Arena C, De Santo AV. Organic matter, nutrient content and biological activity in burned and unburned soils of a mediterranean Maquis area of southern Italy. *Int J Wildland Fire* 2005;**14**:365–77.
114. Girona-Garcia A, Badía-Villas D, Marti-Dalmau C, Ortiz-Perpina O, Mora JL, Armas-Herrera C. Effects of prescribed fire for pasture managment on soil organic matter and biological properties: a 1-year study case in Central Pyrenees. *Sci Total Environ* 2018; **618**:1079–87.
115. Duguy B, Rovira P. Differential thermogravimetry and differential scanning calorimetry of soilorganic matter in mineral horizons. Effects of wildfires and land use. *Org Geochem* 2010;**41**:742–52.
116. Certini G, Nocentini C, Knicker H, Arfaioli P, Rumpel C. Wildfire effects on soil organic matter quantity and quality in two fire-prone Mediterranean pine forests. *Geoderma* 2011;**167**:148–55.
117. González-Pérez JA, González-Vila FJ, Almendros G, Knicker H. The effect of fire on soil organic matter-a review. *Environ Int* 2004;**30**:855–70.
118. Jimenez-Morillo NT, de la Rosa JM, Waggoner D, Almendros G, González-Vila FJ, González-Pérez JA. Fire effects on the molecular structure of soil organic matter fractions under Quercus suber cover. *Catena* 2016;**145**:266–73.
119. Merino A, Chavez-Vergara B, Salgado J, Fonturbel MT, Garcia-Oliva F, Vega JA. Variability in the composition of charred litter generated by wildfire in different ecosystems. *Catena* 2015;**133**:52–63.

120. Mastrolonardo G, Francioso O, Di Foggia M, Bonora S, Forte C, Certini G. Soil pyrogenic organic matter characterization by spetroscopic analysis: a study on combustion and pyrolysis residues. *J Soils Sediments* 2015;**15**:769–80.
121. Alexis MA, Rasse DP, Knicker H, Anquetil C, Rumpel C. Evolution of soil organic matter afer a prescribed fire: a 20 years chronosequence. *Geoderma* 2012; **189–190**:98–107.
122. Santín C, Knicker H, Fernandez S, Menendez-Duarte R, Alvarez MA. Wildfires influence soil organic matter in an Atlantic mountainous region. *Catena* 2008;**74**: 286–95.
123. Faria SR, De la Rosa JM, Knicker H, González-Pérez JA, Keiser JJ. Molecular characterization of wildfire impacts on organic matter in eroded sediments and topsoil in Mediterranean eucalypt stands. *Catena* 2015;**135**:29–37.
124. Deleted in review.
125. Santos F, Russel D, Berhe AA. Thermal alteration of water extractable organic matter in climosequence soils from the Sierra Nevada, California. *J Geophys Res-Bio* 2016;**121**: 2877–85.
126. Nathe K, Levia DF, Tischer A, Michalzik B. Low-intensity Surface fire effects on carbon and nitrogen cycling in soil and soil solution of a Scots pine forest in central Germany. *Catena* 2018;**162**:360–75.
127. Hobley EU, Zoor LC, Shrestha HR, Bennett LT, Weston CJ, Baker TG. Prescribed fire affects the concentration and aromacity of soluble organic matter in forest soils. *Geoderma* 2019;**341**:138–47.
128. Wang Y, Liu X, Yan Q, Hu Y. Impacts of slash and burning on soil carbón pools vary with slope position in a pine plantation in subtropical China. *Catena* 2019;**183**:104212.
129. San Emerito L, Mugica L, Ugarte MD, Goicoa T, Canals RM. Sustainability of radiational pastoral fires in highlands under global change: effects on soil function and nutrient cycling. *Agric Ecosyst Environ* 2016;**235**:155–63.
130. Pottasht K, Meyer S, Crecelius AC, Schubert US, Tisher A, Michalzik B. Land-uses and fire drive temporal patterns of soil solution chemistry and nutrient fluxes. *Sci Total Environ* 2017;**605–606**:514–26.
131. Zhang M, Wang W, Wang D, Heenan M, Xu Z. Short term responses of soil nitrogen mineralization, nitrification and denitrification to prescribed burning in a suburban forest ecosystem of subtropical Australia. *Sci Total Environ* 2018;**642**:879–86.
132. Olefeldt D, Turetsky MR, Blodau C. Altered composition and microbial versus UV-Mediated degradation of disolved organic matter in Boreal Soils following wildfire. *Ecosystems* 2013:1396–412.
133. Wang Q, Zhong M, Wang S. A Meta analysis on the response of microbial biomass, disolved organic matter, respiration and N mineralization in mineral soil to fire in forest ecosystems. *For Ecol Manag* 2012;**271**:91–7.
134. Vergnoux A, Di Rocco R, Domeizel M, Guiliano M, Doumenq P, Theraulaz F. Effects of forest fires on water extractable organic matter and humic substances from Mediterranean soils. UV-vis and fluorescence spectroscopy approach. *Geoderma* 2011;**160**:434–43.
135. Escudey M, Arancibia-Miranda N, Pizarro C, Antilen M. Effect of ash from forest fires on leaching in Volcanic soils. *Catena* 2015;**135**:383–92.
136. Caldwell TG, Johnson DW, Miller WW, Qualls RG. Forest floor carbon and nitrogen losses due to prescription fire. *Soil Sci Soc Am J* 2002;**66**:262–7.
137. DeLuca TH, Zouhar KL. Effects of selection harvest and prescribed fire on the soil nitrogen status of ponderosa pine forests. *For Ecol Manag* 2000;**138**:263–71.
138. Mugica L, Canals R, San Emerito L. Changes in soil nitrogen dynamics caused by prescribed fires in dense gorse lands in SW Pyrenees. *Sci Total Environ* 2018;**639**:175–85.

139. Fultz LM, Moore-Kucera J, Dathe J, Davinic M, Perry G, Wester D, Schwilk DW, Rideout-Hanzak S. Forest wildfire and grassland prescribed fire effects on soil biogeochemical process and microbial communities: two case studies in the semi-arid Southwest. *Appl Soil Ecol* 2016;**99**:118–28.
140. Gimeno-García E, Andreu V, Rubio JL. Changes in organic matter, nitrogen, phosphorous and cations in soil as a result of fire and water erosion in Mediterranean landscape. *Eur J Soil Sci* 2000;**51**:201–10.
141. Pereira-Silva EFL, Casals P, Sodek L, Delitti WBC, Vallejo VR. Post-fire nitrogen uptake and allocation by two resprouting herbaceous species with contrasting belowground traits. *Environ Exp Bot* 2019;**159**:157–67.
142. Hinojosa MB, Parra A, Laudicina VA, Moreno JM. Post-fire functionality and microbial community structure in Mediterranean shrubland subjected to experimental drought. *Sci Total Environ* 2018;**573**:1178–89.
143. Rodriguez J, González-Pérez JA, Turmero A, Hernandez M, Ball AS, González-Vila FJ, Enriqueta Arias M. Wildfire effects on the microbial activity and diversity in a Mediterranean forest soil. *Catena* 2017;**158**:82–8.
144. Barcenas-Moreno G, Garcia-Orenes F, Mataix-Solera J, Mataix-Beneyto J. Plant community influence on soil microbial response after a wildfire in Sierra Nevada National Park. *Sci Total Environ* 2016;**573**:1265–74.
145. St John TV, Rundel PW. The role of fire as a mineralizing agent in a Sierran coniferous forest. *Oecologia* 1976;**25**:35–45.
146. Zavala LM, De Celis R, Jordán A. How wildfires affect soil properties. A brief review. *Cuadernos de Investigacion Geografica* 2014;**40**:311–31.
147. Hatten J, Zabowski D, Sherer G, Dolan E. A comparision of soil properties after contemporary wildfire and fire suppression. *For Ecol Manag* 2005;**220**:227–41.
148. Ludwig B, Khanna PK, Raison RJ, Jacobsen KL. Modelling catión composition of soil extracts under ashbeds after an intense slashfire in a Eucalypt forest. *For Ecol Manag* 1998;**103**:9–20.
149. Larroulet MS, Hepper EN, Redondo MPA, Belmonte V, Urioste AM. The cadenal ecosystem: effects of a prescribed fire on soil chemical properties. *Arid Land Res Manag* 2016;**30**:105–19.
150. Murphy JD, Johnson DW, Miller WW, Walker RF, Carroll EF, Blank RR. Wildfire effects on soil nutrients and leaching in a Tahoe basin watershed. *J Environ Qual* 2006; **35**:479–89.
151. Smith JE, Kluber LA, Jennings TN, McKay D, Brenner G, Sulzman EW. Does the presence of large down Wood at the time of a forest fire impact soil recovery. *For Eco Manage* 2017;**391**:52–62.
152. Kelly R, Montgomery WI, Reid N. Differences in soil chemistry following wildfires on a temperate heath and blanket bog sites of conservation concern. *Geoderma* 2018; **315**:20–6.
153. García-Marco S, González-Prieto S. Short-and medium-term effects of fire and fire-fighting chemicals on soil micronutrient availability. *Sci Total Environ* 2008;**407**: 297–303.
154. Ponder Jr F, Tadros M, Loewenstein EF. Microbial properties and soil nutrients after two prescribed fires in developing savannas in an upland Missouri Ozark Forest. *For Ecol Manag* 2009;**257**:755–63.
155. Lawal M, Marafa LM, Chau KC. Effect of hill fire on upland soil in Hong Kong. *For Ecol Manag* 1999;**120**:97–104.
156. Abraham J, Dowling K, Florentine S. Effects of prescribed fire and post-fire rainfall on mercury mobilization and subsequent contamination assessment in legacy mine site in Australia. *Chemosphere* 2018a;**180**:144–53.

157. Abraham J, Dowling K, Florentine S. Influence of controled burning on the mobility and temporal variations of potential toxic metals (PTMs) in soils of a legacy gold mine site in Central Victoria, Australia. *Geoderma* 2018b;**33**:1–14.
158. Costa MR, Calvao AR, Aranha J. Linking wildfire effects on soils and wáter chemistry of the Marao river watershed, Portugal, and biomass changes detected from Landsat imagery. *Appl Geochem* 2014;**44**:93–102.
159. Campos I, Abrantes N, Keizer JJ, Vale C, Pereira P. Major and trace elements in soils and ashes of eucalypt and pine forest plantations in Portugal following a wildfire. *Sci Total Environ* 2016;**572**:1362–76.
160. Abraham J, Dowling K, Florentine S. The unquantified risk of post-fire metal concentration in soil: a review. *Water Air Soil Pollut* 2017;**228**:175.
161. Yu Q, Rengel Z. Micronutrient deficiency influences plant growth and activities of superoxide dismutases in Narrow-leafed Lupins. *Ann Bot* 1999;**83**:175–82.
162. Li RX, Cai F, Pang G, Shen QR, Li R, Chen W. Solubilization of phosphate and micronutrients by *Trichoderma harzianum* and its relationship with the promotion of Tomato plant growth. *PLoS One* 2015. https://doi.org/10.1371/journal.pone.0130081.
163. Cole JC, Smith MW, Penn CJ, Cheary BS, Conagham KJ. Nitrogen, phosphorous, calcium, and magnesium applied individually or as a slow reléase or controled reléase fertilizer increase growth and yield and affect macronutrient and micronutrient concentration and content of field-grown tomato plants. *Sci Hortic* 2016;**211**:420–30.
164. Penn CJ, Camberato JJ. A critical review on soil chemical processes that control how soil pH affects phosphorous availability to plants. *Agriculture* 2019;**9**:120.
165. Schaller J, Ticher A, Struyf E, Bremer M, Belmonte DU, Potthast K. Fire enhances phosphorous availability in topsoils depending on binding properties. *Ecology* 2015;**96**:1598–606.
166. Butler OM, Elser JJ, Lewis T, MacKey B, Chen C. The phosphorous rich signature of fire in the soil-plant system: a global meta-analysis. *Ecol Lett* 2018;**21**:335–44.
167. Santín C, Otero X, Doerr SH, Chafer CJ. Impact of a moderate/high severity prescribed eucalypt forest fire on soi phosphorous stocks and partitioning. *Sci Total Environ* 2018;**621**:1103–14.
168. Rau BM, Blank RR, Chambers JC, Johnson DW. Prescribed fire in a Great Basin sagebrush ecosystem: dynamics of extractable nitrogen and phosphorous. *J Arid Environ* 2007;**71**:362–75.
169. Wang G, Yu X, Bao K, Xing W, Gao G, Lin Q, Lu X. Effect of fire on phosphorous forms in sphagnum moss and peat soils of ombrotrophic bogs. *Chemosphere* 2015;**119**:1329–34.
170. Garcia-Oliva F, Merino A, Fonturbel MT, Omil B, Fernandez C, Vega JA. Severe wildfire hinders renewal of soil P pools by termal mineralization of organic P in forest soil: analysis by sequential extraction and ^{31}P NMR spectroscopy. *Geoderma* 2018;**309**:32–40.
171. Guenon R, Vennetier M, Dupuy N, Roussos S, Pailler A, Gros. Trends in recovery of Mediterranean soil chemical properties and microbial activities after infrequent and frequent wildfires. *Land Degrad Dev* 2013;**24**:115–28.
172. Pellegrini A, Ahlstrom A, Hobbie SE, Reich PB, Nieradzik LP, Staver C, Scharenbroch BC, Jumpponen A, Anderegg WRL, Randerson JT, Jackson RB. Fire frequency drives decadal changes in soil carbon and nitrogen and ecosystem productivity. *Nature* 2018;**553**:194–8.
173. Lombao A, Barreiro A, Cancelo-Gonzalez J, Martin A, Diaz-Ravina M. Impact of termal shock on soils affected by fires of different severity. *Spanish J Soil Sci* 2015;**5**:165–79.

174. Karavani A, Boer MM, Baudena M, Colinas C, Diaz-Sierra R, Peman J, De Luis M, Enriquez-de-Salamanca A, De Dios VR. Fire-induced deforestation in drought-prone Mediterranean forests: drivers and unknowns from leaves to communities. *Ecol Monogr* 2018;**88**:141–69.
175. Batllori E, De Caceres M, Brotons L, Ackerly DD, Moritz MA, Lloret F. Cumulative effects of fire and drought in Mediterranean Ecosystems. *Ecosphere* 2017;**8**:e01906.
176. Turner MG, Braziunas KH, Hansen WD, Harvey BJ. Short-interval severe fire erodes the resilience of subalpine lodgepole pine forests. *Proc Natl Acad Sci USA* 2019;**116**: 11319–28.
177. Wittenberg L, Tessler N. Ash and soils. A close relationship in fire-affected areas. In: Pereira P, Mataix-Solera J, Úbeda X, Rein G, Cerdà A, editors. *Fire effects on soil properties*. CSIRO Publishing (Australia); 2019. p. 185–201.
178. Muqaddas B, Chen C, Lewis T, Wild C. Temporal dynamics of carbon and nitrogen in surface soil and forest floor under different prescribed burning regimes. *For Ecol Manag* 2016;**382**:110–9.
179. Muqaddas B, Zhou X, Lewis T, Wild C, Chen C. Long-term frequent prescribed fire decreases surface soil carbon and nitrogen pools in wet sclerophyll forest of Southeast Queensland, Australia. *Sci Total Environ* 2015;**536**:39–47.
180. Wright RJ, Hart SC. Nitrogen and phosphorous status in a ponderosa pine forest after 20 years of interval burning. *Ecoscience* 1997;**4**:526–33.
181. Williams RJ, Hallgren SW, Wilson GWT. Frequency of prescribed burning in an upland oak forest determines soil litter properties and litter properties and alters the soil microbial community. *For Ecol Manag* 2012;**265**:241–7.
182. Binkley D, Ritcher D, David MB, Caldwell B. Soil chemistry in a loblolly/longleaf pine forest with interval burning. *Ecol Appl* 1992;**2**:157–64.
183. Gonzalez-Pelayo O, Andreu V, Gimeno-Garcia E, Rubio JL. Rainfall influence on plot-scale runoff and soil los from repeated burning in a Mediterranean-shrub ecosystem, Valencia, Spain. *Geomorphology* 2010;**118**:444–52.
184. Taylor QA, Midgley MG. Prescription side effects. Long-term high frequency controled burning enhances nitrogen availability in an Illinois oak-dominated forest. *For Ecol Manag* 2018;**411**:82–9.
185. Coates TA, Hagan DL, Aust WM, Johnson A, Keen JC, Chow AT, Dozier JH. Mineral and soil chemical properties as influenced by long-term use of prescribed fire with differing frequencies in a Southeastern coastal plain pine forest. *Forests* 2018;**9**:739.
186. Cheng CH, Chen YS, Huang YH, Chiou CR, Lin CC, Menyailo OV. Effects of repeated fires on ecosystem C and N stocks along a fire induced forest/grassland gradient. *J Geophys Res-Bio* 2013;**118**:215–25.
187. Mayor AG, Valdecantos A, Vallejo VR, Keizer JJ, Bloem J, Baeza J, Gonzalez-Pelayo O, Machado AI, de Ruiter PC. Fire-induced pine Woodland to shrubland transitions in Southern Europe may promote shifts in soil fertility. *Sci Total Environ* 2016;**573**:1232–41.
188. Keesstra S, Wittenberg L, Maroulis J, Sambalino F, Malkinson D, Cerdà A, Pereira P. The influence of fire history, plant species and post-fire management on soil water repellency in a Mediterranean catchment: the Mount Carmel range. *Catena* 2017; **149**:827–66.
189. Hosseini M, Keizer JJ, Gonzalez-Pelayo O, Prats SA, Ritsema C, Geissen V. Effects of fire frequency on runoff, soil erosion, and loss of organic matter at the micro-plot scale in northcentral Portugal. *Geoderma* 2016;**269**:126–37.
190. Hosseini M, Geissen V, Gonzalez-Pelayo O, Serpa D, Machado AI, Ritsema C, Keizer JJ. Effects of fire occurrence and recurrence on nitrogen pools and phosphorous losses by overland flow in maritime pine plantations in north-central Portugal. *Geoderma* 2017;**289**:97–106.

191. Kinner DA, Moody JA. Spatial variability of steady-state infiltration into a two-layer system on burned hillslopes. *J Hydrol* 2010;**381**:322–32.
192. Moody JA, Shakesby RA, Robichaud PR, Cannon SH, Martin DA. Current research issues related to post-wildfire runoff and erosion processes. *Earth Sci Rev* 2013;**122**:10–37.
193. Cerdà A. Changes in overland flow and infiltration after a rangeland fire in a Mediterranean scrubland. *Hydrol Process* 1998;**12**:1031–42.
194. Gimeno-García E, Andreu V, Rubio JL. Influence of vegetation recovery on water erosion at short and medium-term after experimental fires in a Mediterranean shrubland. *Catena* 2007;**69**:150–60.
195. Woods SW, Balfour V. The effect of ash on runoff and erosion after a forest wildfire, Montana, U.S.A. *Int J Wildland Fire* 2008;**17**:535–48.
196. Cerdà A, Doerr SH. The effect of ash and needle cover on surface runoff and erosion in the immediate post-fire period. *Catena* 2008;**74**:256–63.
197. Zavala LM, Jordán A, Gil J, Bellinfante N, Pain C. Intact ash and charred litter reduces susceptibility to rain splash erosion post-wildfire. *Earth Surf Process Landforms* 2009;**34**: 1522–32.
198. Ebel BA, Moody JA, Martin DA. Hydrologic conditions controlling runoff generation immediately after wildfire. *Water Resour Res* 2012;**48**:W03529. https://doi.org/10.1029/2011WR011470.
199. Bodí MB, Doerr SH, Cerdà A, Mataix-Solera J. Hydrological effects of a layer of vegetation ash on underlying wettable and water repellent soil. *Geoderma* 2012;**191**:14–23.
200. Larsen IJ, MacDonald LH, Brown E, Rough D, Welsh MJ, Pietraszek JH, Libohova Z, Benavides-Solorio JD, Schaffrath K. Causes of post-fire runoff and erosion: water repellency, cover, or soil sealing? *Soil Sci Soc Am J* 2009;**73**:1393–407.
201. Arcenegui V, Mataix-Solera J, Guerrero C, Zornoza R, Mayoral AM, Morales J. Factors controlling the water repellency induced by fire in calcareous Mediterranean forest soils. *Eur J Soil Sci* 2007;**58**:1254–9.
202. Dekker LW, Ritsema CJ. Soil hydrophobicity variations with depth and particle size fraction in burned and unburned *Eucalyptus globulus* and *Pinus pinaster* forest terrain in the Águeda basin, Portugal. *Catena* 1996;**27**:25–47.
203. Hallet PD, Young IM. Changes to water repellence of soil aggregates caused by substrate induced microbial activity. *Eur J Soil Sci* 1999;**50**:35–40.
204. McGhie DA, Posner AM. The effect of plant top material on the water repellence of fired sands and water repellent soils. *Aust J Agric Res* 1981;**32**:609–20.
205. Jordán A, Zavala LM, Mataix-Solera J, Doerr SH. Soil water repellency: origin, assessment and geomorphological consequences. *Catena* 2013;**108**:1–5.
206. Jordán A, Martínez-Zavala L, Bellinfante N. Heterogeneity in soil hydrological response from different land cover types in southern Spain. *Catena* 2000;**74**:137–43.
207. Mataix-Solera J, Doerr SH. Hydrophobicity and aggregate stability in calcareous topsoils from fire-affected pine forests in southeastern Spain. *Geoderma* 2004;**118**:77–88.
208. DeBano LF, Mann LD, Hamilton DA. Translocation of hydrophobic substances into soil by burning organic litter. *Soil Sci Soc Am J* 1970;**34**:130–3.
209. Badía D, Martí C, Aguirre AJ, Aznar JM, Gonzalez-Pérez JA, De la Rosa JM, León J, Ibarra P, Echevarria T. Wildfire effects on nutrients and organic carbon of a Rendzic Phaeozem in NE Spain: changes at cm-scale topsoil. *Catena* 2014;**113**:267–75.
210. Soto B, Benito E, Díaz-Fierros F. Heat-induced degradation processes in forest soils. *Int J Wildland Fire* 1991;**1**:147–52.
211. DeBano LF. The role of fire and soil heating on water repellency in wildland environments: a review. *J Hydrol* 2000;**231–232**:195–206.
212. Martin DA, Moody JA. Comparison of soil infiltration rates in burned and unburned mountainous watersheds. *Hydrol Process* 2001;**15**:2893–903.

213. Kunze MD, Stednick JD. Streamflow and suspended sediment yield following the 2000 Bobcat Fire, Colorado. *Hydrol Process* 2006;**20**:1661–81.
214. Bento-Gonçalves A, Vieira A, Úbeda X, Martin D. Fire and soils: key concepts and recent advances. *Geoderma* 2012;**191**:3–13.
215. Moody JA, Martin DA. Synthesis of sediment yields after wildland fire in different rainfall regimes in the western United States. *Int J Wildland Fire* 2009;**18**:96–115.
216. Nyman P, Sheridan GJ, Smith HG, Lane PNJ. Evidence of debris flow occurrence after wildfire in upland catchments of south-east Australia. *Geomorphology* 2011;**125**: 383–401.
217. Sheridan G, Lane P, Nyman P. Erosion. In: Pereira P, Mataix-Solera J, Úbeda X, Rein G, Cerdà A, editors. *Fire effects on soil properties*. CSIRO Publishing (Australia); 2019. p. 89–113.
218. Wondzell SM, King JG. Postfire erosional processes in the Pacific Northwest and Rocky Mountains regions. *For Ecol Manag* 2009;**178**:75–87.
219. Cawson JG, Sheridan GJ, Smith HG, Land PNJ. Surface runoff and erosion after prescribed burning and the effect of different fire regimes in forests and shrublands: a review. *Int J Wildland Fire* 2012;**21**:857–72.
220. Pereira P, Cerdà A, Úbeda X, Mataix-Solera J, Arcenegui V, Zavala L. Modelling the impacts of wildfire on ash thickness in a short term period. *Land Degrad Dev* 2015;**26**: 180–92.
221. Mayor AG. Post-fire hydrological and erosional responses of a Mediterranean landscape: seven years of catchment-scale dynamics. *Catena* 2007;**71**:68–75.
222. Poesen J, Lavee H. Rock fragments in top soils: significance and processes. *Catena* 1994;**23**:1–28.
223. Ruiz-Sinoga JD, Romero-Díaz A, Ferre E, Martínez-Murillo JF. The role of soil surface conditions in regulating runoff and erosion processes on a metamorphic hillslope (Southern Spain). Soil surface conditions, runoff and erosion in Southern Spain. *Catena* 2010;**80**:131–9.
224. Gordillo-Rivero AJ, García-Moreno J, Jordán A, Zavala LM, Granja-Martins FM. Fire severity and surface rock fragments cause patchy distribution of soil water repellency and infiltration rates after burning. *Hydrol Process* 2014;**28**:5823–43.
225. Cerdà A, Bodí MB. Erosión hídrica en suelos afectados por incendios forestales. In: Mataix-Solera J, editor. *Incendios forestales, suelos y erosión hídrica, (coord.)*. Edit. Caja de Ahorros del Mediterráneo CEMACAM: Alcoi; 2007. p. 71–118.
226. Scott DF, Curran MP, Robichaud PR, Wagenbrenner JW. Soil Erosion after forest fire. In: Cerdà A, Robichaud P. (Eds.), Fire effects on soils and restoration strategies. Science Publishers, pp. 177-195.
227. Martínez-Murillo JF, Hueso-González P, Ruiz-Sinoga JD, Lavee H. Short-term experimental fire effects in soil and water losses in southern Spain. *Land Degrad Dev* 2016;**27**:1513–22.
228. Fernández C, Vega JA, Jiménez E, Fonturbel T. Effectiveness of three post-fire treatments at reducing soil erosion in Galicia (NW Spain). *Int J Wildland Fire* 2011; **20**:104–14.
229. Ruiz-Gallardo JR, Castaño S, Valdés A. Cartografiado de la prioridad de actuación postincendio forestal. In: Cerdà A, Mataix-Solera J, editors. *Efectos de los incendios forestales sobre los suelos en España. El estado de la cuestión visto por los científicos españoles, Càtedra de Divulgació de la Ciència*. Universitat de Valencia; 2009. p. 445–67.
230. Martínez-Murillo J, López-Vicente M. Effect of salvage logging and check dams on simulated hydrological connectivity in a Burned area. *Land Degrad Dev* 2018;**29**: 701–2.

231. Miller HJ, Yool SR. Mapping forest post-fire canopy consumption in several overstory types using multi-temporal Landsat TM and ETM data. *Remote Sens Environ* 2002;**82**: 481–96.
232. Van Wagtendonk JW, Root RR, Key CH. Comparison of AVIRIS and Landsat ETM+ detection capabilities for burn severity. *Remote Sens Environ* 2004;**92**:397–408.
233. Chuvieco E, Riaño D, Danson FM, Martín P. Use of a radiative transfer model to simulate the postfire spectral response to burn severity. *J Geophyis Res-Bio* 2006;**111**: G04S09.
234. De Santis A, Chuvieco E, Vaughan PJ. Short-term assessment of burn severity using the inversion of PROSPECT and GeoSail models. *Remote Sens Environ* 2009;**113**: 126–36.
235. Patterson MW, Yool SR. Mapping fire-induced vegetation mortality using Landsat thematic mapper data: a comparison of linear transformation techniques. *Remote Sens Environ* 1998;**65**:132–42.
236. Brewer CK, Winne JC, Redmond RL, Opitz DW, Mangrich MV. Classifying and mapping wildfire severity: a comparison of methods. *Photogramm Eng Remote Sens* 2005;**71**:1311–20.
237. Moreno JM, Oechel WC. A Simple Method for estimating fire intensity after a burn in California Chaparral. *Acta Ecol* 1989;**10**:57–68.
238. Pérez B, Moreno JM. Methods for quantifying fire severity in shrubland fires. *Plant Ecol* 1998;**139**:91–101.
239. Wang GG. Fire severity in relation to canopy composition within burned boreal mixewood stands. *For Ecol Manag* 2002;**163**:85–92.
240. Cocke AE, Fule PZ, Crouse JE. Comparison of burn severity assessments using Differenced Normalized Burn Ratio and ground data. *Int J Wildland Fire* 2005;**14**:189–98.
241. Epting J, Verbyla DL, Sorbel B. Evaluation of remotely senses indices for assessing burn severity in interior Alaska using Landsat TM and ETM+. *Remote Sens Environ* 2005;**96**: 328–39.
242. Hyde K, Woods SW, Donahue J. Predicting gully rejuvenation after wildfire using remotely sensed burn severity data. *Geomorphology* 2007;**86**:496–511.
243. Kokaly RF, Rockwell BW, Haire SL, King TVV. Characterization of postfire surface cover, soils, and burn severity at the Cerro Grande Fire, New Mexico, using hyperspectral and multispectral remote sensing. *Remote Sens Environ* 2007;**106**:305–25.
244. Miller JD, Thode AE. Quantifying burn severity in a heterogeneous landscape with a relative version of the delta Normalized Burn Ratio (dNBR). *Remote Sens Environ* 2007;**109**:66–80.
245. Ferreira AJD, Coelho COA, Ritsema CJ, Boulet AK, Keizer JJ. Soil and water degradation processes in burned areas: lessons learned from a nested approach. *Catena* 2008; **74**:273–85.
246. Prats SA, MacDonald LH, Monteiro MSV, Ferreira AJD, Coelho COA, Keizer JJ. Effectiveness of forest residue "mulching" in reducing post-fire runoff and erosion in a pine and a eucalypt plantation in north-central Portugal. *Geoderma* 2012;**191**: 115–24.
247. Neary DG, Klopatek CC, DeBano LF, Ffolliott PF. Fire effects on belowground sustainability: a review and synthesis. *For Ecol Manag* 1999;**122**:51–71.
248. MacDonald LH, Larsen IJ. Effects of forest fires and post-fire rehabilitation: a Colorado case study. In: Cerdà A, Robichaud PR, editors. *Fire effects on soils and restoration strategies*. Science Publishers; 2009. p. 423–52.
249. Robichaud PR, Lewis SA, Wagenbrenner JW, Ashmun LE, Brown RE. Post-fire mulching for runoff and erosion mitigation. Part I: effectiveness at reducing hillslope erosion rates. *Catena* 2013;**105**:75–92.

250. Bautista S, Robichaud PR, Bladé C. Post-fire mulching. In: Cerdà A, Robichaud PR, editors. *Fire effects on soils and restoration strategies*. Science Publishers; 2009. p. 353–72.
251. Bautista S, Bellot J, Vallejo VR. "Mulching" treatment for post-fire soil conservation in a semiarid ecosystem. *Arid Soil Res Rehabil* 1996;**10**:235–342.
252. Prats SA, Martins MAS, Cortizo-Malvar MA, Ben-Hur M, Keizer JJ. Polyacrylamide application versus forest residue mulching for reducing post-fire runoff and soil erosion. *Sci Total Environ* 2014;**468–469**:464–74.
253. Pannkuk CD, Robichaud P. Effectiveness of needle cast at reducing erosion after forest fires. *Water Resour Res* 2003;**32**:1333.
254. Deleted in review.
255. Groen AH, Woods SW. Effectiveness of aerial seeding and straw mulch for reducing post-wildfire erosion, north-western Montana, USA. *Int J Wildland Fire* 2008;**17**:559–71.
256. Yanosek KA, Foltz RB, Dooley JH. Performance assessment of wood strand erosion control materials among varying slopes, soil textures and cover amounts. *J Soil Water Conserv* 2006;**61**:45–51.
257. Smets T, Poesen J, Knapen A. Spatial scale effects on the effectiveness of organic mulches in reducing soil erosion by water. *Earth Sci Rev* 2008;**89**:1–12.
258. Foltz RB, Dooley JH. Comparison of erosion reduction between wood strands and agricultural straw. *Trans ASAE (Am Soc Agric Eng)* 2003;**46**:1389–96.
259. Foltz RB, Wagenbrenner NS. An evaluation of three wood shred blends for post-fire erosion control using indoor simulated rain events on small plots. *Catena* 2010;**80**:86–94.
260. Foltz RB, Copeland NS. Evaluating the efficacy of wood shreds for mitigating erosion. *J Environ Manag* 2009;**90**:779–85.
261. Theng BKG. Clay–polymer interaction: summary and perspectives. *Clay Clay Miner* 1982;**30**:1–10.
262. Vacher CA, Loch RJ, Raiste SR. Effect of polyacrylamide additions on infiltration and erosion of disturbed lands. *Aust J Soil Res* 2003;**41**:1509–20.
263. Sojka RE, Bjorneberg DL, Entry JA, Lentz RD, Orts WJ. Polyacrylamide in agriculture and environmental land management. *Adv Agron* 2007;**92**:75–162.
264. Ben-Hur M. Using synthetic polymers as soil conditioners to control runoff and soil loss in arid regions—a review. *Aust J Soil Res* 2006;**44**:191–204.

CHAPTER FIVE

Mining environments

María Gabarrón, Raúl Zornoza*, Jose A. Acosta, Ángel Faz and Silvia Martínez-Martínez
Sustainable Use, Management and Reclamation of Soil and Water Research Group, Department of Agricultural Engineering, Universidad Politécnica de Cartagena, Cartagena, Colombia
*Corresponding author: E-mail: raul.zornoza@upct.es

Contents

Advances in Chemical Pollution, Environmental Management and Protection, Volume 4
ISSN 2468-9289
https://doi.org/10.1016/bs.apmp.2019.07.003

Abstract

The mining industry is an economic driver in several countries worldwide but it also represents an important source of environmental problems. The main impacts caused by the metallic and non-metallic mining activity are the loss of vegetation cover and the physical, chemical and biological degradation of soils, which promote alterations in the ecosystem, landscape and watercourses. To repair the ecosystem services and productivity of these areas, remediation technologies based on physicochemical techniques, bioremediation (use of microorganisms) or phytotechnologies (use of vegetation associated to the symbiotic microorganisms) should be carried out. Furthermore, good management of mining wastes during the exploitation period can help to prevent soil degradation in surrounding areas as well in the tailings ponds themselves.

Keywords: Mining activity; Landscape; Vegetation cover; Soil degradation; Reclamation; Bioremediation; Phytotechnologies; Waste management.

1. Introduction

Mining may well have been the second of humankind's earliest endeavors— granted that agriculture was the first. The two industries are ranked together as the primary or basic industries of early civilization. The importance of these industries has changed very little since the beginning of civilization. If we consider fishing and lumbering as part of agriculture and oil and gas production as part of mining, then agriculture and mining continue to supply all the basic resources used by modern civilizations.[1]

Mining has played an important part in human existence from prehistoric times right through to the present.[2] The history of mining is fascinating and parallels the history of civilization, with many important cultural eras associated with and identified by various minerals or their derivatives: the Stone Age (prior to 4000 BCE), the Bronze Age (4000—5000 BCE), the Iron Age (1500 BCE to CE 1780), the Steel Age (1780—1945), and the Nuclear Age (1945 to the present).

Nowadays, mining is first and foremost a source of mineral commodities which all countries find essential for maintaining and improving their standards of living. Mined materials are needed in order to construct roads and hospitals, to build automobiles and houses, to make computers and satellites, to generate electricity, and to provide the many other goods and services that consumers enjoy. In addition, mining is economically important for the producing regions and countries. It provides employment, dividends, and taxes that pay for hospitals, schools, and public facilities. The mining industry produces a trained workforce and small businesses that can service communities and may also initiate related businesses. Mining also yields foreign exchange and accounts for a significant portion of gross domestic product. Mining fosters a number of associated activities, such as the manufacturing of mining equipment, the provision of engineering and environmental services, and the development of world-class universities in the fields of geology, mining engineering, and metallurgy. The economic opportunities and wealth generated by mining are substantial for many producing countries.[3]

According to data provided by Seri and Wu,[4] over the last decades, the extraction of mineral resources increased globally in order to meet the market requirements. However, in Europe the data indicate an almost constant activity in the last three decades in terms of tonnage: ~5 000 Mt (million tonnes) of run-of-mine were excavated every year in Europe. Therefore, while the worldwide extraction increased by almost 200% over the last three decades (from 22 280 Mt (106 t) in 1980 to 61 730 Mt in 2013), the European contribution at global level dropped from ~23% in 1980 to ~ 8% in 2013. These data show the current importance of the mining industries, which provide us with most of the materials we rely on to build infrastructures and instruments for daily use, to obtain large amounts of energy, and to supply agriculture with fertilizers that enable most foods to be produced.

At the same time, mining is the human activity that has most disturbed the environment and is linked to large social impacts and inequalities. Notwithstanding this, our future is deeply dependent on mining. Current mining practices require good management of mining wastes and processes to change and contribute to community development with greater equity, and to better protect natural resources and ecosystems in order to be environmentally acceptable and compliant with sustainable development objectives. Additionally, current and legacy mining sites should be reclaimed in order to reestablish their natural status, using all the reclamation techniques available to avoid the impacts that the mining activity promotes in soils, vegetation and landscape.[5]

1.1 Mining activity

Mining is the process of excavating ore minerals along with minimum waste rocks from the Earth's crust for the benefit of humankind. Before commenting on the existing types of mining process and activities, it is necessary to clarify the following terms in the field of mining in order to avoid confusion among them and to be able to adequately understand the different mining processes.[6]

- *Mine:* an excavation made in the earth to extract minerals.
- *Mining engineering:* the planning and design of mines, considering economic, technical, and geologic factors; also, supervision of the extraction, and sometimes the preliminary refinement, of the raw material.

  On the other hand, there are two main types of mining:
- *Metallic mining:* is the prowess of the means to haul out metal deposits from underneath the crust of the earth, for the metals that are deposited at the top of the earth's crust are very little or merely do not exist. Hauling out of the metal deposits happens with the lot of novel techniques that keeps brimming up every now and then. Metallic mining constitutes the extraction activity to obtain a specific metal. The metallic elements are classified into four types:
  - Basic: Copper, lead, zinc, tin.
  - Ferrous: Iron, manganese, molybdenum, cobalt, tungsten, titanium, chromium.
  - Precious: Gold, silver, platinum.
  - Radioactive agents: plutonium, uranium, radium, thorium.
- *Non-metallic mining:* is the extraction of stone, sand, rock or similar materials from natural deposits. The most common examples of non-metallic mines are quarries and pits.

Some terms distinguish various types of mined minerals. Geologically, one can distinguish the following mineral categories:

- *Mineral:* a naturally occurring inorganic element or compound having an orderly internal structure and a characteristic chemical composition, crystal form, and physical properties.

  The most common classification of minerals includes metallic, non-metallic, and energy minerals. The metallic minerals are subclassified as (1) ferrous (iron, manganese, nickel, and cobalt), (2) non-ferrous (copper, zinc, lead, tin, and bauxite), and (3) precious (gold, silver, and platinum).[7] The non-metallic minerals are composed of quartz, feldspar, mica, garnet, potash, sulfur, and salt. The energy minerals contain coal, petroleum, and natural gas.
- *Rock:* An aggregate of one or more minerals, e.g., granite, shale, marble; or a body of undifferentiated mineral matter, e.g., obsidian, or of solid organic material, e.g., coal.

Minerals and rocks can be classified into various formats based on the type and chemistry of minerals, geographic distribution, depth of occurrence, morphology, relation to host rocks, nature of mineralization, structural control, genetic model, economic gradation fitting with the overall characteristics, and perspective. A particular mineral or rock that will be formed from a certain combination of elements depends on the physical and chemical conditions under which the material forms. This, in turn, results in a wide range of color, hardness, crystal forms, luster, and density that a particular mineral possesses.[8]

The economic differences in the nature of mineral deposits are evident in the following terms:

- *Ore:* the naturally occurring material from which a mineral or minerals of economic value can be extracted profitably or to satisfy social or political objectives. The term is generally, but not always, used to refer to metalliferous material, and is often modified by the names of the valuable constituent; e.g., iron ore.; ore mineral.
- *Gangue:* The valueless minerals in an ore; that part of an ore that is not economically desirable but cannot be avoided in mining. It is separated from the ore minerals during concentration.
- *Waste:* the part of an ore deposit that is too low in grade to be of economic value at the time of mining, but which may be stored separately for possible treatment later. Gangue is a particular type of waste.

Next, the main mining activities will be explained: handling lose ground, drilling and blasting of hard rocks, removal of broken materials from the workplace, and supporting the ground for safe operations. Various mining methods are available to exploit different types of deposits.[6,9,10] The prime objective is to mine in the safest conditions and economically without sacrificing the interest of the conservation of minerals -a non-renewable wasting asset. The choice of mining techniques depends on the following[11]: a) Nearness to surface; b) Nature of overburden; c) Shape, size, regularity, and continuity; d) Strike, dip, thickness, and rock strength; e) Nature of mineralization; f) Host and wall rock condition; g) Stripping/overburden and ore-to-waste removal ratio; h) Possibility of minimizing internal and external dilution; i) Availability of infrastructures; j) Cost of mining and mineral dressing; k) Production target and resource/reserve status; and l) Value of primary, associated commodities and value added elements.

The overall sequence of activities in modern mining is often compared with the five stages in the life of a mine: prospecting, exploration, development, exploitation, and reclamation.

1.1.1 Prospecting

The first stage in the utilization of a mineral deposit is the search for ores or other valuable minerals (coal or non-metallic). Because mineral deposits may be located either at or below the surface of the earth, both direct and indirect prospecting techniques are employed. The direct method of discovery, normally limited to surface deposits, consists in visual examination of either the exposure (outcrop) of the deposit or the loose fragments (float) that have weathered away from the outcrop. Geologic studies of the entire area augment this simple, direct technique. By means of aerial photography, geologic maps, and structural assessment of an area, the geologist gathers evidence by direct methods to locate mineral deposits. Precise mapping and structural analysis plus microscopic studies of samples also enable the geologist to locate hidden as well as surface mineralization.

The most valuable scientific tool employed in the indirect search for hidden mineral deposits is geophysics, the science of detecting anomalies using physical measurements of gravitational, seismic, magnetic, electrical, electromagnetic, and radiometric variables of the earth. The methods are applied from the air, using aircraft and satellites; on the surface of the earth; and beneath the earth, using methods that probe below the topography. Geochemistry, the quantitative analysis of soil, rock, and water samples, and geobotany, the analysis of plant growth patterns, can also be employed as prospecting tools.[6]

1.1.2 Exploration

The second stage in the life of a mine, exploration, determines the size and value of a mineral deposit as accurately as possible, utilizing techniques similar to, but more refined than, those used in prospecting. The demarcation line between prospecting and exploration is somewhat blurred; in fact, in some cases a distinction may not be possible. Exploration generally shifts to surface and subsurface locations, using a variety of measurements to obtain a more positive picture of the extent and grade of the ore body. Representative samples may be subjected to chemical, metallurgical, X ray, spectrographic, or radiometric evaluation techniques that are meant to enhance the investigator's knowledge of the mineral deposit. Samples are obtained by chipping outcrops, trenching, tunneling, and drilling; in addition, borehole logs may be provided to study the geologic and structural makeup of the deposit. Rotary, percussion, or diamond drills can be used for exploration purposes. However, diamond drills are favored because the cores they yield provide knowledge of the geologic structure. These cores are normally then split along their axis; one-half is analyzed, while the other half is retained intact for further geologic study.[6]

An evaluation of the samples enables the geologist or mining engineer to calculate the tonnage and grade, or richness, of the mineral deposit. Technical personnel estimate the mining costs, evaluate the recovery of the valuable minerals, determine the environmental costs, and assess other foreseeable factors in an attempt to reach a conclusion regarding the profitability of the mineral deposit. The key question is to determine whether the property is just another mineral deposit or an ore body. For an ore deposit, the overall process is called reserve estimation, that is to say, the examination and valuation of the ore body. Once this stage is concluded, the project is developed, traded to another party, or abandoned.[12]

1.1.3 Development

If the mineral ore exploration phase proves that there is a large enough mineral ore deposit, of sufficient grade, then the project proposer may begin to plan for the development of the mine.

This phase of the mining project has several distinct components:[13]

a) *Construction of access roads:* the construction of access roads, either to provide heavy equipment and supplies to the mine site or to ship out processed metals and ores, can have substantial environmental impacts, especially if access roads cut through ecologically sensitive areas or are near previously isolated communities. If a proposed mining project involves the construction of any access roads, then the environmental impact assessment for the project must include a comprehensive assessment of the environmental and social impacts of these roads.

b) *Site preparation and clearing:* If a mine site is located in a remote, undeveloped area, the project proponent may need to begin by clearing land for the construction of staging areas that would house project personnel and equipment. Even before any land is mined, activities associated with site preparation and clearing can have significant environmental impacts, especially if they are within or adjacent to ecologically sensitive areas. The environmental impact assessment must evaluate, separately, the impacts associated with site preparation and clearing.

1.1.4 Exploitation

This is the fourth stage of mining, and is associated with the actual recovery of minerals from the earth in quantity. Although development may continue, the emphasis in the exploitation stage is on production. Prior to exploitation there is usually only enough development to ensure that production, once started, can continue uninterrupted throughout the life of the mine. The mining method selected for exploitation is determined mainly by the characteristics of the mineral deposit and the limits imposed by safety, technology, environmental concerns, and economics. Geologic conditions, such as the dip, shape, and strength of the ore and the surrounding rock, play a key role in selecting the method.[14]

When the ore is extracted from the mine, it needs to be ground (or milled) and the relatively small quantities of metal are separated from the non-metallic material of the ore in a process called "beneficiation." Beneficiation includes physical and/or chemical separation techniques such as gravity concentration, magnetic separation, electrostatic separation, flotation, solvent extraction, electrowinning, leaching, precipitation, and amalgamation (often involving the use of mercury). Leaching involving the use of cyanide is a kind of beneficiation process, usually employed with gold, silver, and copper ores, which merits separate attention because of the serious environmental and public safety impacts. With leaching, finely ground ore is deposited in a large pile (called a "leach pile") on top of an impermeable pad, and a solution containing cyanide is sprayed on top of the pile. The cyanide solution dissolves the desired metals and the "pregnant" solution containing the metal is collected from the bottom of the pile using a system of pipes.[15]

1.1.5 Reclamation

When active mining ceases, mine facilities and the site are reclaimed and closed. The goal of mine site reclamation and closure should always be to return the site to a condition that most resembles the pre-mining condition. Mines that are notorious for their immense impact on the environment often only made impacts during the closure phase, when active mining operations ceased. These impacts can persist for decades and even centuries. Therefore, the environmental impact assessment for every proposed mining project must include a detailed discussion of the mine Reclamation and Closure Plan offered by the mining proponent. Mine reclamation and closure plans must describe in sufficient detail how the mining company will restore the site to a condition that most resembles pre-mining environmental quality; how it will prevent — in perpetuity — the release of toxic contaminants from various mine facilities (such as abandoned open pits and tailings impoundments); and how funds will be set aside to ensure that the costs of reclamation and closure will be paid for. Unfortunately, the legislation requiring the restoration of mining sites is relatively new in many countries, which is why there are numerous mining areas worldwide that have not been reclaimed and that currently generate environmental and health problems.[16]

1.2 Mining methods

There are two basic ways in which minerals are mined (extracted) – surface or open cast mining and underground mining. The choice of extraction method depends on the local topography; depth and type of mineral being mined; the shape, size and location of the ore body; and cost considerations. Solution mining (in-situ leaching) is a specialized, less common process used on certain types of metal and sulfide ore deposits.[17]

1.2.1 Surface or open-cast mining

A surface mine generally consists of a large open-pit dug into the earth or along the side of hill or mountain with very high pit walls. This kind of mine is the least expensive kind, and is every miner's first choice where an ore body is situated close to the surface, is big enough, and has little overburden. Open-cast mines look simple, but every pit needs to be tailor made. First and foremost, the pit walls must be stable and stay up, so it is important to understand the rock-mechanics of the pit walls to determine a safe slope for the pit. There is also a delicate balance between how much overburden and waste rock can be mined in order to gain access to the valuable ore and how deep a pit can be. An open-pit mine is constructed in a series of benches, decreasing in size from the surface to the bottom of the mine. The size and location of the first bench of any open pit mine is critical. It is excavated well into the rock surrounding an ore body. Since each successive bench is smaller than the one that was removed above it, the depth to which the pit can be mined is determined by the size and location of the first cut or bench. Surface mining requires the removal and disposal of a layer of soil and rock containing no minerals, commonly called the overburden. A second layer of rock, known as waste rock, containing low concentrations of ore is also removed and, depending on the ore content, is either processed or disposed. The ratio between the tonnages of waste to ore is known as the stripping ratio. The lower this ratio is the less the waste rock that has to be removed is and so the operating costs are lower. After the overburden and waste rock are removed to expose the ore body, ore is drilled, blasted, loaded into trucks and hauled to the appropriate facility for disposal, beneficiation, or processing. Once the high-quality deposit is exposed, excavation continues, with further disposal of surrounding low-grade waste rock, until the valuable body of ore has been removed. After initial crushing and grinding, the high-quality ore is then transported to the mill for further processing. Lower grade ore may be transported to a heap leach pad where a solution is applied to release the mineral from the rock. The main cost advantage of open-pit mining versus underground is that the miners can use large and more powerful shovels and trucks, because the equipment is not restricted by the size of the opening it must work in. This enables faster production, and the lower cost also means that lower grades of ore can be mined. If an ore body is large, and extends from the surface to a great depth, it is common to start mining near the surface from an open pit. This provides some early revenue while preparations are made for underground mining of the deeper parts of the ore body. It is not uncommon for ore below the floor of an open pit to be developed from underground by driving a ramp (adit) from the lower part of the pit.[18]

1.2.2 Underground mining

Underground mining is appropriate for that part of the ore body where an open pit operation is uneconomic due to a higher ore-to-overburden ratio. The deposit is either deep seated or has significant vertical continuity. The underground method recovers ore existing below the surface safely and economically.[19] The technique generates as little waste as possible, and therefore waste handling and stockpiling are negligible, which makes the activities environmentally friendly. The excavation of steeply inclined thin vein type and massive ore body proceeds either upward by overhand or downward by underhand stopping methods, and is broadly grouped under two categories: a) Soft rock mining: Coal, potash, rock salt; b) Hard rock mining: Copper, lead-zinc, gold, chromite, uranium, and platinum-group elements.

1.2.3 Solution mining

Solution mining or in situ leaching is an alternative to the underground and surface techniques described above. Its use for the mining of metal oxide and sulfide ores has increased since 1975. Solution mining involves drilling and pumping a diluted sulfuric acid or other reagent solution directly into the below-ground ore body. The reagent dissolves the metals in the ore and the solution is collected by various means such as wells or sumps. It is then pumped to the surface and recovered using electrical/chemical techniques. Solution mining has enabled facilities to beneficiate lower-grade sulfide and oxide ores. Solution mining presents unique environmental challenges since it requires constant management of the solutions deep underground.[20]

1.3 Mine wastes

Mining produces large amounts of waste. While the amount of waste can be substantial, not all waste is considered harmful to the environment. Some waste might even be safe to use in other applications, such as in building materials. However, when the rocks contain certain minerals in large enough quantities, combined with poor management of the waste, then it can entail a large risk to the environment and the health of local communities. Waste occurs in several stages of the mining process and throughout the life of the mine, from the first exploration drilling project to the last processed material before the mine's closure. Several types of waste are generated in a mine, but three types stand out as having the largest volume: waste rock, tailings, and mine water. In some cases, the mine uses certain chemicals such as cyanide in the processing stage. These chemicals often account for a small volume of the total waste, but can pose a large risk due to their high toxicity. A summary of the most important types of waste from mines is given below.[21]

1.3.1 Waste rock

This is the material that is excavated to reach the ore. The amount of waste rock that needs to be removed depends on the geometry and location of the ore body, along with the mining method used and the composition and stability of the rocks. The ratio between waste rock and ore production is called the stripping ratio. Waste rock is often stored close to the mine to minimize transport. The waste rock is deposited in piles or heaps. The composition of the waste rock controls what elements can be released to the environment. Some elements are highly toxic even in small concentrations, for example mercury or some compounds of nitrogen such as ammonia, while other elements are less toxic but pose a threat in higher concentrations, for example zinc and copper. There are often different types of waste rock within a mine; some might be more harmful than others. One effective way of working with waste rock management is the separation of different types of waste rock.[22]

1.3.2 Acid mine drainage

When water and air start to break down the waste rock, the process is called weathering, the elements can be released to the environment. A large block of rock weathers slowly as it has a low surface area, while smaller pieces create higher surface areas, making them more susceptible to weathering. Sulfide minerals are easily weathered in contact with oxygen. When sulfide minerals break down, they can produce acid water. The acid water further speeds up the weathering of the minerals, called chemical weathering. The result is acid water with a high metal content, called acid rock drainage (ARD) or acid mine drainage (AMD) when in association with mines.

1.3.3 Tailings

When the ore is processed and the valuable minerals are separated, a fine-grained mineral sand remains as waste, called tailings. The amount of tailings that remains is determined by the grade of the ore; this refers to the percentage of valuable minerals in the ore. For example, if the grade of an ore is 1% copper, 99% of the total ore would be deposited as tailings. Iron ores generally have higher grades than sulfide or gold ores, often exceeding 50% or even more. Fewer tailings are therefore generally produced in iron ore projects.[22]

After separating the valuable minerals, the tailings are often pumped in a slurry through pipes to be deposited in a tailings storage facility (TSF). The pipes are equipped with spigots in several different places around the TSF to evenly distribute the tailings. The TSF is built especially for tailings storage and is designed for the location and the geology. There are different types of TSFs and ways of building. The most common type of TSF is placed on the surface and built with high embankments that are engineered to retain the tailings and water. This type is called conventional impoundment storage.[23]

The embankments of the TSF are often built with waste rock from the mine. The waste rock needs to be non-acid generating rock to ensure long-term stability. The embankments sometimes include a layer of geomembrane, a type of synthetic liner or barrier with a low permeability, which is mainly used to prevent the fine-grained tailings from seeping into the coarse waste rock. The composition of the tailings is similar to the ore. As with waste rock, it is important to characterize the tailings to know what the composition is: which minerals are present, and in what ratios. The tailings are fine-grained, meaning that the surface area is large and weathering is expected to be high. This is especially important when there are sulfide minerals involved, meaning that the chemical weathering of those minerals will be high. To mitigate the oxidation of the sulfide minerals, the tailings can be kept under water. When tailings are held under water it is referred to as a subaqueous deposition. This technique helps to minimize the oxidation of mainly sulfide minerals, but also influences the way the tailings settle in the TSF, which in turn impacts the long-term stability. Subaerial deposition - deposition above water - is more common than subaqueous.[24]

1.3.4 Mine water

Water is used in many of the processes within the mine and the processing plant. Some processes, such as milling, can use large volumes of water and potentially contaminate it with metals and other harmful elements. Water is valuable, and is often recycled and reused over and over again in different processes. Eventually the water is collected and released to the environment. On occasions the water must be treated in different ways prior to being released. Water treatment can be categorized into active or passive treatments.[25]

Active water treatment means that the water is treated within a water treatment facility. Several methods exist for active treatment, but these often include raising the pH of the water with the addition of elements such as limestone or caustic soda, which in turn cause metals to precipitate and settle. This can be done in certain settling or sedimentation ponds. In some cases, flocculants are added to make the process quicker. The treatment can also be complemented with other methods, such as filter systems.[26]

Passive water treatment means that the water is treated without active involvement, sometimes called a "walk-away solution", often using natural physical, chemical or biological processes. Common techniques for passive water treatment involve the use of plants that uptake contaminants or which use soils that filter the contaminants and deposit them in a controlled area.[26]

Passive water treatment is cheaper than active treatment, which is generally a major cost within mine operations. However, treating highly acidic waters with passive techniques remains difficult and therefore the combination of active and passive treatment techniques can be used to lower costs.[26]

1.3.5 Chemicals used in processing

The chemicals used in processing are in a much lower volume than the other types of wastes, but can constitute a major risk due to high toxicity, even in small concentrations. A common chemical used in gold extraction is cyanide. The cyanide dissolves the gold to an aqueous solution, making it a simple and cost-effective technique. Cyanide is highly toxic for both people as well as for the environment. Even low doses of cyanide can cause permanent damage in humans or animals. However, aqueous cyanide is easily hydrolyzed, especially when in contact with UV-rays from the sun. When the cyanide is hydrolyzed it breaks down to other chemical compounds that are less toxic. The cyanide solution is kept in a closed system during the extraction phase and is then hydrolyzed, sometimes in a so-called cyanide destruction plant.[27]

The large volumes of waste produced at mining operations are expensive to manage, and are frequently cited as an obstacle in the environmental sustainability of mining.

The scarcity of resources and increased environmental focus has contributed to increasing awareness around the reuse of these wastes. The mining activity plays a leading role in waste management, and is one of the few industries that recycles its own waste. Some uses of mine waste include:[28]

a) *Manganese tailings*: have been used in agro-forestry, building and construction materials, coating, glass or resin among others.
b) *Sulfur oxide emissions*: many smelters have installed acid plants to convert sulfur dioxide to sulfuric acid, a useful industrial chemical.
c) *Clay-rich tailings*: have been used for making floor tiles, bricks and cement.
d) *Waste rock*: can be reprocessed to extract minerals and metals, used as back fill, landscaping material, feedstock for cement and concrete, or aggregate in road construction.
e) *Slag*: is often used for road construction, and in concrete and cement.
f) *Mine water*: is used for dust suppression and mineral processing, as a source of drinking water and agriculture uses, as a coolant.
g) *Red mud*: Bauxite red mud is solid alkaline waste produced in aluminum refineries. Red mud has been used as a raw material for ceramics, glass, and bricks and as a soil amender.
h) *Water treatment sludge*: sludge from acid rock drainage treatment, which is high in iron, has been sold commercially for use in pigments.

2. Impact of mining on landscape and vegetation

Metallic and non-metallic mining activity often results in large areas being disturbed by changes in the land surface and ecosystem disruption. The landscape is mainly perceived as a visual resource with an esthetic function, although it also has an ecological one. However, it is more difficult to define and assess the esthetic function than the ecological function.[29]

The landscape pattern is defined as the arrangement and combination of a series of patches with different shapes and sizes in the spatial domain.[30] The alteration of some of these patches in active and post-mining areas results in a negative visual impact of the mining sites on the surroundings, with the adjacent landscape being degraded by the negative visual impact of the unreclaimed sites and the infrastructures associated to the mine activity such as roads, bridges, conveyor belts or large industrial complexes.[31]

The alteration of the landscape pattern is especially harmful when the mine activity is developed in National Parks or areas of natural beauty, firstly for the visual disruption of the scene and finally and more importantly by the habitat fragmentation and loss of biodiversity in areas of especial interest for their wildlife.

Open-cast mining is, perhaps, the most aggressive and most directly perceivable type of mining. It drastically transforms the relief; removing large volumes of soil, changing the soil and vegetation cover of the mined area and altering water streams.[32,33] Fig. 1 shows a copper open-cast mine in Bor (Serbia).

Fig. 1 Copper open-cast mine in Bor (Serbia) (top) and tailings and tailings ponds from metallic abandoned mines in Spain (two bottom rows).

The visual impact of underground mining is less shocking than that of open-cast mining as the galleries cannot be seen, but the construction of mining structures to raise extracted materials to the surface and soil surfaces broken by subsidence also disturb the natural scenery. Underground mining is also linked to several environmental impacts such as structural instability and soil subsidence, which modify water streams and vegetation cover.[34]

Both open-cast and underground mining extracted minerals undergo separation treatments to obtain the valuable part. The commercially valueless materials conform the waste material, known as mine spoils, which is dumped out of the mine area.

In coal and metallic mining, these mine spoils are usually stacked in tailings and tailings ponds, conforming new reliefs that alter the landscape pattern (Fig. 1).

Besides the strictly visual alteration, the physical and chemical properties of metallic and non-metallic mine spoils contribute to the fragmentation of the landscape because they suppose an aggravating factor to the loss of vegetation cover caused by the soil removal, conditioning the establishment of plants in unreclaimed mining areas. Promoting a good vegetation cover in degraded lands, such as mining sites, is very effective in reducing surface erosion because the roots bind the substrate. Also, vegetation promotes the establishment of microorganism communities and can return a large proportion of percolating water to the atmosphere through transpiration, thus reducing the concentrations of soluble heavy metals entering watercourses. Certain key factors linked to mine spoils properties limit plant growth: the substrate structure, the lack of macronutrients and high levels of metals.[35]

The grain size of the residues determines restrictions for plant establishment because of the water retention capacity and the compaction and even cementation processes that make surfaces impenetrable for plants.[35]

The lack of organic matter limits plant growth because the necessary contribution of the nutrients that it carries is not produced. In addition, organic matter helps to improve the soil structure by increasing porosity and aggregate formation, which in turn improves water retention and micronutrients. Likewise, the lack of clay minerals, which provide cation exchange sites in normal soils aiding to retain inorganic nutrients provided by the organic matter, results in a leaching of nutrients.

Therefore, amelioration of the structure and organic matter content of the mining waste must be a priority in reclamation processes of tailings ponds, for vegetation growth.[36]

In metalliferous residues the most inhibitory factor to plants growth is the metal toxicity caused by the presence of high amounts of metals in a form that enables plant uptake. The availability of metals refers to the fraction of metals in the soil solution that is readily available for migration to other soil compartments, leaching through the soil/tailing profile or uptake by biota.[37] Mechanisms of phytotoxicity are poorly understood and strongly specific for each species and metal, but it is known that high amounts of metals and metalloids such as Cu, Cr, Pb, Ni, Zn, or As, result in growth restriction, changes in cellular structures and functions or susceptibility to drought.[36,38]

An increase in the risk of metals plant uptake is related to their total content in the mining wastes, although its mobility is strongly related to other physicochemical properties such as the forms of occurrence of metals (sulfides, hydrated anions or cations, hydroxides, oxides ...) or the environmental conditions in the tailings pond such as pH or redox stratus.[36,37] The combined effects of pH and redox potential (Eh) on the mobility of trace elements are complex and highly specific for each element. Nevertheless, it is known that acidic conditions on tailings ponds cause an increase in the solubility of Fe, Al and other toxic elements due to the decrease in sorption sites where these hydrated cations could be adsorbed. So, metals stay as free cations in the soil solution, available to be uptaken by plant roots. Conversely, elements that exist as anions, for example As, Mo, Se, Cr (VI), are more mobile in alkaline conditions.[37]

Metals tend to be less mobile in reducing conditions than in oxidizing conditions, since sulfides are likely to control the behavior of several elements in reducing conditions.[181] In this state, most sulfides are in the form of precipitated salts, making metals more unavailable for plants. Changes from reducing to oxidizing conditions involve transformations of sulfides and a shift to more acidic conditions, involving periodically high levels of trace elements in the soil pore water.

The oxidation of metal sulfides in mining wastes, mainly pyrite, is a major cause of acidification in these environments. Upon oxidation of a reduced soil or sediment, the important components that buffer against a decrease in pH are carbonates, exchangeable cations, clays and Al hydroxides.[37] This buffering activity is extremely important in reclamation processes. Other minor factors have deleterious effects on vegetation, including extreme temperatures, wind erosion or slope instability.

The importance of developing vegetation cover goes a long way toward reducing the visual scars in the landscape caused by large-scale mining operations. Successful revegetation may enable recreational use of the land, and even agriculture or forestry if conditions are favorable. A well-planned revegetation scheme should overcome the problems on a permanent basis. Several models have been developed to predict the phytoavailability of trace elements, especially of Cd, Zn, Cu, and Pb, but tend to be limited to a given plant and specific growth conditions.[39] Hence, this requires a thorough site evaluation and the selection of the most appropriate revegetation technique with regard to local conditions such as geology, climate or toxicity levels.[35]

3. Impact of mining on soil

Soils are amongst the most important natural resources, defining sustainable development and independence of states. In recent decades, the geochemical characteristics of soils have undergone significant changes under the influence of anthropogenic activities worldwide.[40] On the other hand, the mining industry is both a leading factor of economic development in a country but also a significant source of environmental problems.[41–43] In fact, the largest impacts related to mining activity have generally occurred in soils.

During excavation, transport and dumping, the original soil structure and properties are drastically altered. The rearranged soil is disturbed by mining operations and can be defined as mine soils or Technosols,[44–47]which develop from the mixture of fragmented rock and fine earthy material.[48] Thus, the quality and properties of mine soil vary among locations depending on local conditions (e.g. geology, climate, land-use).[49] As an important part of the mining system, the mine soil system plays an important role in regulating different subsystems of the mining area, such as plant, water, and landscape subsystems. In fact, mine soil quality largely determines the future orientation of reclamation.

The physical, chemical and biological properties of mine soils are highly degraded. These properties are interconnected and play important roles in the development of a functioning soil system on severely disturbed lands.[50] Any interference with the connection will lead to the degradation of mine soil. Over many centuries, soils have developed a sophisticated self-regulation capacity involving nutrient supply, chemical buffering, soil aggregate stabilization, porosity, aeration, water holding capacity, detoxification and self-regeneration.[51] This self-regulation capacity is driven by the five soil formation factors (i.e. climate, organisms, topography, parent material, and time) that tie the soil and other systems together.[52] However, mining often causes drastic changes in the soil profile, altering the equilibrium among the five above-mentioned factors, and resulting in a loss of soil biodiversity, soil organic carbon, nutrient pools, toxicity, etc.[53] As a direct result, the soil system can suffer substantial and long-lasting damage, which negatively affects land productivity and ecosystem functionality.[54]

Therefore, mining activity results in structural and functional changes in mine soils, such as a high level of soil compaction, acidity, and toxicity; limitations in soil water holding capacity and the availability of nutrients; and the imbalance of soil microbial processes. All of these changes can be clustered into three types of degradation: physical, chemical and biological.

3.1 Soil physical degradation

Since mining uses heavy machinery and involves blasting during mineral extraction, important soil properties are altered such as the disruption of stable soil aggregates, thus decreasing the porosity and aeration, increasing the bulk density, and eventually depriving the soil of organic matter and nutrients, also affecting organisms (meso and microfauna). These soils, or newly created substrates or technosols, are often inhospitable to vegetation due to a combination of physical, chemical and microbiological factors.[55]

The main factor affecting soil physical degradation is the use of heavy equipment for ore extraction and transport. Highly mechanized mining processes are widely used in surface and subsurface mining because they generally provide an efficient approach for achieving high production.[56] However, high production is also associated with high soil physical degradation and large volumes of waste production.[57] During mining, the excavated materials from deep layers of subsoil are stripped off and vary substantially in their physical and chemical properties, such as soil bulk density, water holding capacity and water absorbing capacity.[58] Therefore, surface materials such as technosols and rock spoils on the post-mining landscape can vary widely in their physical consistency and suitability for plant species.[45] These physical differences manifest in mine soil structural and functional changes. Physical properties highly altered by mining, such as soil texture, bulk density, and aeration have a large influence on soil remediation by controlling soil hydraulic properties and hydrological stability.

One of the most common physical degradations on mine sites is soil compaction. Soil compaction is the physical form of soil degradation that changes soil structure and productivity,[59] and compaction poses a potential threat to sustainable development in mining and reclamation practices. In a mining area, the most frequently used parameter to characterize mine soil compaction is bulk density.[60] Due to the use of heavy machinery during mining and reclamation processes, mine soils often have higher bulk densities (1.55−1.86 Mg m-3).[61] In compacted soils, the air and water exchange capacity is directly affected. Thus, for an accurate measurement of mine soil compaction, other properties, such as soil porosity, strength and moisture, must be evaluated.[62] Compacted mine soils prevent seed germination and the establishment of native vegetation.

Another important physical degradation of mine soil is soil erosion, both wind and water erosion, which are widely known. The mobilization and the transport of soil particles by wind or water primarily causes losses of the fine and less dense particle fractions, including humus particles and clay, both of which are important carriers of soil nutrients and stabilizing agents for the physical properties of soils. In addition, erosion from mine-waste dumps and roads is another environmental problem of mine areas, where polluted particles can reach populated areas or agricultural fields, with the subsequent associated human health risk. Both natural (air and water flows, temperature changes, migration of pore solutions, electrolytes) and technogenic factors (for example, impacts of heavy mining equipment and various mechanisms) influence erosion intensity, as well as internal factors such as the physical and chemical properties, and the composition and structure of depositions. Human health and environmental risks from soils are also associated to contaminated soil resulting from windblown dust. Fugitive dust can pose significant environmental problems at some mines.[63] The inherent toxicity of the dust depends upon the proximity of environmental receptors and the type of ore being mined. High levels of As, Pb, and radionuclides in windblown dust commonly pose the greatest risk.

3.2 Soil chemical degradation

The effects of mining on certain soil chemical properties have been investigated by several researchers e.g. Shrestha and Lal[56] or Zhen et al.[64] Chemical degradation of soil is often associated to soil pollution by heavy metals; however, this type of degradation involves many other processes that negatively affect soil functions, such us salinization, acidification and alkalization, depletion of nutrients and loss of organic matter.[65]

Mining can contaminate soils by metals over a large area, especially agricultural and urban areas near a mining project may be particularly affected. Heavy metals and metalloids from mine activities include Pb, Cd, Cr, Hg, As, Zn, etc. When excess amounts of heavy metals enter the pedosphere, soil quality decreases due to the reduction of soil productivity.[66] In addition, heavy metals within the mine residues and acid drainage travel into the water and soil and begin to circulate among these spheres through weathering and dissolution processes,[67] changing the mine soil chemical properties. Moreover, soils contaminated from chemical spills and residues at mine sites may pose a direct contact human risk when these materials are misused as fill materials, in ornamental landscaping, or as soil supplements.[68]

Another chemical degradation commonly found in mine areas concerns the drastic changes in soil pH. Low pH is a particularly intransigent problem in mine wastes that contain iron pyrites, which, upon weathering, will generate sulfuric acid and (if there is no acidic neutralizing capacity in the waste) induce pH values of <2.0. The oxidation of sulfidic waste rocks results in extreme acidity, high salinity (mainly SO42−) and elevated levels of metals in mine areas. Martínez-Martínez et al.[69] reported that the remnants and tailings from extraction and processing of ores consisting of galena, sphalerite, and pyrite lead to the formation of a low pH and high contents of soluble metals. In contrast, soil pH can also increase in the mining areas where alkaline materials are processed, especially surrounding magnesite calcination plants, where the pH can increase up to maximum values of 9.5.[70] This soil alkalization has negative effects on the function of soil microbes and the phytoavailability of nutrient elements in soil solution.[71]

Although soil is highly sensitive to mining activities, it plays a crucial role in material cycling in the pedosphere through balancing soil nutrient, water and energy flows that ensure productivity and sustain biodiversity. However, when mining removes the vegetation and inevitably leads to the loss of some plant nutrients from the site, the nutrient balance is highly altered.[72] Assel[73] found a lower and reduced content of soil N and P in mine soils when compared to the natural forest, which was due to ecosystem disruption, removal of vegetation and loss of litter layer during mineral mining. In addition, Assel[73] found a SOC content on a mined area to be 0.14%, which is below the accepted level of SOC of soil fertility, giving an indication of the disruption of ecosystem functioning and a loss of organic matter due to mineral extraction and processing.

In addition, nearly all mine soils have very low levels of macro-nutrients, especially N, P and K. Hilson[74] argued that regardless of the overburden type used, plant available N and P tend to be low in mined soils, which may limit plant growth. Sheoran et al.[75] also added that the consequences of physical disturbance to the topsoil during stripping, stockpiling and reinstatement, cause unusually large N transformations and movements eventually leading to a substantial loss in soil fertility and productivity.

Amegbey[72] also indicated that as topsoil is only about 20 cm deep, and contains most of the plants' available nutrients, the scraping action with bulldozers and other heavy machinery depletes the soils of the fertility and productivity, exposing the subsoil, which is unsuitable for plant growth. Furthermore, stockpiling of top soil in mounds, which is a very common phenomenon during mineral extraction, has been shown to affect the biological, chemical and physical properties of the soil.

3.3 Soil biological degradation

Soil has an important role in maintaining the biodiversity of mine sites. Doran and Parkin[76] defined soil quality as the capacity of the soil to maintain the environment, improve plant and animal health and function properly within the ecosystem to provide biological productivity. As has been mentioned, mining activities represent the source of potentially toxic metal pollution, which has a detrimental effect on soil quality, particularly affecting soil biodiversity.

Soil biota govern processes related to nutrient, energy and organic matter cycling. The biodiversity of soil fauna reflects the ecosystem metabolism; it is therefore often used as an indicator of soil quality to evaluate the recovery conditions in mining areas. The abundance of soil fauna is related to microhabitat conditions, which reflect the distribution of organic matter as an important factor that affects microclimate.[77]

Many studies have been conducted in mining areas, mainly focusing on the changes in biomass of microfauna and mesofauna and the activity of soil enzymes. These studies contain not only the changes in biological properties but also the spatial variation of biota caused by dumping and overburdened topography. Moreover, the diversity and dynamic complexity of animals, plants and microorganism communities constitutes the ecosystem.[78,79,183]

Organisms such as mesofauna and macrofauna (e.g., microarthropods, macroarthropods, enchytraeids and earthworms) are only a minor fraction of the total living biomass in soil,[80] but the appropriate amount of these organisms also reflects the condition of mine soils. Larger animals are called mesofauna and include organisms such as earthworms, arthropods, and large nematodes; macrofauna also include burrowing mammals, such as moles and rabbits, which can be indicators of the condition of the soil at mine sites. The most affected mine soils have poorly developed mesofauna communities.

Microorganisms are mainly affected by the loss of organic matter and nutrients depletion. The removal of topsoil during surface mining instantly reduces the pool of soil organic matter and therefore decreases the amount of microorganisms and reduces its biodiversity[81]; additionally, the habitat of soil organisms is greatly altered. As the microbial community is sensitive to the soil environment and is related to diverse soil processes, including decomposition of organic residues, nutrient cycling, and degradation of toxic compounds and pollutants,[82] it has been used as an ecological indicator in severely disturbed mine sites by many researchers.[83–86]

In addition, soil enzymes are an important component of soil quality. They regulate the function of the ecosystem and play key biochemical functions in the overall process of organic matter transformation and nutrient cycling in the soil system.[87,88] Therefore, they are also considered as an ecological indicator of soil quality.[89,90] Various intracellular and extracellular enzymes that originate from microorganisms, plants, and animals constitute the overall enzyme activity in soil. Enzymatic activity can be easily influenced by the disturbance of mine soils. This is due to soil enzymes being very sensitive to natural and anthropogenic disturbances, and thus responding very quickly to increased metal concentrations.[90–93] So, a study of enzyme activity in soil provides useful information on the extent of the environmental impact of toxic metals. Hu et al.[90] reported that declined activity of urease was very high, which is followed by dehydrogenase and acid phosphatase in comparison to that of control soil due to greater concentrations of Cu, Zn, and Cd in soil affected by mining activities. It was observed that the reductions in the percentage of enzymatic activity in comparison to control soil were 58%, 40%, 24%, 27%, and 37% for urease, acid phosphatase, alkaline phosphatase, β-glucosidase, and dehydrogenase, respectively. Since urease is the most sensitive enzyme studied in soil, it can be regarded as the indicator of soil environmental quality.[94]

4. Soil reclamation techniques

Soil reclamation is defined as the general process of repairing disturbed, damaged, degraded, or destroyed land with respect to its former or future productive use. Thus, reclamation is a broad concept encompassing all other terms commonly referring to soil restoration, rehabilitation or replacement, used alone or combined.[112] Restoration means to restore a mine site to the previously existing conditions; rehabilitation indicates reparation of ecosystem services and productivity, while replacement results in an

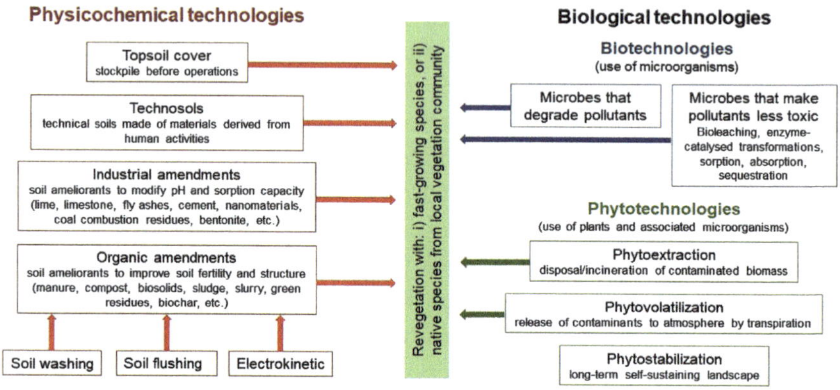

Fig. 2 Overview of mine soil reclamation strategies.

alternative land use, aiming at its economic and esthetic enhancement, aligned with rural development goals.[95]

The goal of mine reclamation is therefore, not necessarily to restore the pre-disturbance condition, but at least to reduce the hazards to the environment and public health and to improve the esthetic value of the area. Owing to the high impact of mining activity on the soil, reclamation strategies are mandatory in many countries to avoid soil pollution, biodiversity loss and landscape deterioration. A major challenge in mine soil reclamation is the re-establishment of a self-sustaining native vegetation cover. For this purpose, microorganisms or vegetation have been used in some cases to reduce the toxicity of toxic mine compounds, owing to their ability to decontaminate the areas or reduce the toxicity of the contaminants. However, using a standardized mix of low diversity exotic species has negatively impacted on local flora and fauna by resulting in lower plant and animal diversity at local and regional scales.[96] In the following subsections we explain in detail the most used strategies followed for reclaiming mine soils based on physicochemical techniques, bioremediation (use of microorganisms) or phytotechnologies (use of vegetation associated to the symbiotic microorganisms), which is also summarized in Fig. 2.

4.1 Physicochemical techniques

The addition of topsoil on the upper surface of the affected area (normally stockpiled before mining operations, others brought from external regions) and the use of soil amendments are the most widely used techniques for mine soil reclamation worldwide, with important concerns for erosion control and minimizing, in many cases, acid mine drainage.[96] In metalliferous mine areas where ores are related to sulfide minerals, pH control is critical to reduce soil and water pollution by toxic elements and to promote the development of a self-sustaining environment though activation of primary production. The best technique used to modify tailings pH is the use of proper amendments, which normally contribute not only to modify pH, by also to reduce the toxicity of toxic elements by precipitation processes and increases in sorption capacity. Several amendments, such as lime, calcium carbonate (as limestone, marble residues or animal shells) and fly ash, have been successfully applied for mine soil reclamation, with positive long-term effects on pH control and the immobilization of toxic metals.[97,98,103,106,155,179,180]

All manner of organic amendments have been used for reclamation purposes, such as manures, composts, biosolids, sewage sludge, animal slurries, agroindustry residues, paper mill residues, plant residues, biochar, etc.[97,99–103] It is well known that organic matter facilitates the development of a stable vegetation community by improving the soil quality, by increases in soil fertility, soil aggregation, soil porosity, water retention, micro- and macro-organisms biomass, diversity and activity. Thus, organic amendments are used to recover the lost productivity of mining environments. Antonelli et al.[104] observed that a single addition of biosolids in tailings of a former metalliferous mining area in Canada promoted long-term carbon sequestration and ameliorated soil properties necessary to sustain primary production. For coal mine reclamation, the use of coal combustion residues has proved effective and economically feasible. These residues include fly ash, gypsum and sulfite rich FGD materials composed of lime, fly ash and filter cake.[105]

Another efficient strategy to reclaim mine soils is the use of Technosols. A Technosol is defined as a soil developed from materials made or derived from human activity (i.e. waste, mine tailings).[182] So, instead of incorporating amendments in the soil, the affected mine area is directly covered with a technical soil made for the purpose of reclamation. The adequate choice of those materials is critical for soil reclamation, since they have to transform harmful compounds into less toxic forms, so microorganisms, vegetation and animals can grow, and primary production is boosted. In the case of metalliferous mines, this goal can be achieved by the adoption of materials with metal stabilization potential, to transform bioavailable metal species into geochemically stable forms.[106] A variety of stabilizing materials have been used, such as limestone, marble waste, lime, cement, glacial till, mussel shell, nanomaterials (zeolites, metal oxides, carbon nanotubes, noble metals, and titanium dioxide) peat, manures, compost, bentonite or biochar.[102,107–109]

Carbonates, bentonite and nanomaterials in Technosols, for example, have been reported to have a high capacity to immobilize soil metals by changes in pH and by high specific sorption.[99,106,108] Nanomaterials have the major characteristic of having a high surface-to-volume ratio, and so the sorption capacity of toxic elements is greater than that of other materials. In addition, owing to their small size, they can move through soil pores and so be more effective at adsorbing toxic compounds.[108,110]

Under these approaches, the surface of the disturbed mine area is covered by a soil (in the case of adding topsoils or Technosols) or by geological materials mixed with industrial and organic amendments so that a soil is developed with time. The main aim of these strategies is to achieve a soil with enough quality to promote primary production, so that permanent vegetation cover is achieved. In some cases, no sowing or planting has been performed, and native plant species from the surroundings have been able to colonize the reclaimed area and start an ecological succession process in the area.[100,109,111,180] However, this process is slow, and is mostly dependent on climatic conditions, seed reservoirs and soil physical, chemical and biological properties. Thus, the most frequent practice after the addition of soil or amendments is revegetation, either with fast-growing plant species to reduce the surface of bare soil and reduce erosion rates, or by native species from the climatic vegetation community of the area.[96,99,103,107,112]

When the aim of the reclamation strategy requires the removal of the present contaminants and not just a decrease in their mobility or toxicity, other physical methods are due. Electrokinetic soil reclamation is an in-situ technology that removes ionic metal contaminants from soils under the influence of an applied direct current by electrodes in saturated soil, resulting in transport by electroosmosis and ionic migration.[113,114] Electro-osmosis mobilizes the pore fluid to flush the soil system, usually toward the cathode, while ionic migration effectively separates the negative and positive ionic species by their transport to the anode and cathode, respectively.[113] This technique is able to remove metals that are in the soil as ionic forms, in dissolution or retained by soil colloids by sorption mechanisms. The attraction of metal ions to the soil colloids depends primarily on the soil electronegativity and the dissociation energy of ions.[114] Thus, desorption of ionic metals from colloids is essential to remove metals from contaminated soils using electrokinetics. The attraction of metal ions to soil is also dependent on pH. Thus, the higher the content of organic matter, clays and carbonates, the lower the metal removal efficiency owing to desorption resistance. The electric current generated during the process induces reactions on the electrodes and in the polluted matrix that help in the dissolution of metal ions, which are transported and concentrated by migration. Therefore, the operating conditions, such as the initial pH of the soil and intensity of the electric field may be adjusted to favor the dissolution and transportation of metals out of the solids matrix.[115] Metal contaminants concentrated at the polarized electrodes are removed by electroplating, (co-) precipitation, solution pumping, or ion exchange resin complexation.[116] Depending on the overall velocity at which metal ions move in the soil, electrokinetic reclamation may take from a few days to several years.[113,114,116] To improve the effectiveness of this technique, different complexing agents such as citric acid, maleic acid, DTPA, EDTA or nitrilotriacetic acid have been applied to the soil to facilitate the desorption of metals to soil colloids.[117] This technology has been effective at decreasing the availability of Ad, Cu, Mn, Cr, Hg, Pb, and Zn,[115,118–122] reducing the toxicity of the soil, which could then be naturally colonized by the surrounding vegetation or revegetated by sowing/planting.

Soil washing is another technology used ex situ to remove the content of available metals from soils. This technique is used when there is an urgent need to clean up the soil by contaminants. Soil washing involves the removal of metals and metalloids from the soil using various reagents and extractants.[123] To proceed with this technique, the soil must be dug out and mixed with an extractant solution, selected in terms of the specific characteristics of the metal/metalloid and the soil. Then, the soil/extractant mix is agitated for a specific time, and through ion exchange, chelation, adsorption or precipitation, the metals or metalloids in the soil matrix are transferred to the liquid. The clean soil can then be backfilled to the original site.[116,123] Effective results have been found for many reagents tested for soil washing, these include: DTPA, EDTA, EDDS, formic acid, tartaric acid, acetic acid, citric acid, oxalic acid, ammonium chloride, ammonium acetate, hydrochloric acid, phosphoric acid, etc.[123–126] The in-situ variant of soil washing is the technology called soil flushing. Instead of removing the soil from the area, an extraction liquid with reagents/extractants is made to pass through the soil.[127–130] The extraction fluid rich in metals and metalloids removed from the soil matrix by ion exchange, chelation, or adsorption is recovered, reused or disposed of. Soil flushing is mostly applicable to homogenous, coarse-textured soils with high permeability.[131]

The drawback of electrokinetics, soil washing and soil flushing is that all metal ions in the soil are removed, including essential nutrients. Thus, the soil becomes clean of contaminants but also lacks the nutrients to foster primary production. As a consequence, after the application of these techniques, the addition of organic amendments is encouraged to recover soil fertility and to promote the development of vegetation and microbial communities.

4.2 Biotechnologies

The use of biotechnologies implies the use of microorganisms to reclaim mine soils, taking advantage of the biogeochemical cycles, which they are specifically involved in. The aim of biotechnologies is to decompose toxic organic contaminants and to turn toxic inorganic contaminants into less toxic forms at safe levels. The relation between mining and microorganisms dates back thousands of years. In this line, the ancient Romans already observed that a blue color in the liquid of the tailings was associated to the presence of a microorganism (*Thiobacillus ferrooxidans*), with the capacity to transform low quality mine ore into profitable metals such as Cu.[132] Furthermore, other bacterial species such as *Pseudomonas cepacia* and *Erwinia herbicola* are able to remove phosphate from its ore without the presence of sulfuric acid.[133] Microorganisms have acquired the ability to resist the effect of toxic compounds through adaptation to extreme environments such as metalliferous materials with low pH physiological mechanisms and so, survive in those environments which are not hospitable to higher organisms.[132] In recent years, there has been a great effort in engineering microbial species to enhance their ability to reclaim mine soils contaminated with metals. However, the limitation for this strategy is the very little information available on the diversity of microorganisms living in mine soils. The advance in next generation sequencing (NGS) in recent years gives light to the identification of new microbial species with genes involved in the biogeochemical cycle of toxic elements in mining environments, which could be used for soil reclamation; however, research is still in progress.

So, most efforts taken to reclaim mine soils have addressed the addition of overburdens and amendments, and very little work has been performed to date on the potential application of microorganisms, owing to the difficulty to i) identify them in the native environment; ii) isolate and reproduce them; iii) boost their ability to reclaim toxic environments through genetic engineering; iv) achieve soil reclamation in short periods of time; v) handle them under real field conditions, and/or vi) avoid the release of intermediate toxic by-products, in some cases.[134–138]

The microbial biodiversity of mine soils varies greatly among regions, owing to differences in the strategy of mining extraction processes. It varies from Eubacteria, Archaea and Proteobacteria.[132] Most of the species have heavy metal resistance, and are extremophiles tolerating extreme pH and temperatures. Most strategies based on microorganisms to reclaim mine soils rich in metals and metalloids are based on their ability to mobilize or immobilize metals and metalloids in the soil.[134] One of the most efficient genera for this end is *Bacillus*. Microorganisms can reclaim soils by promoting: i) bioleaching, forming soluble metal chelates by the release of organic acids into the environment; ii) enzyme-catalyzed transformations such as redox reactions; iii) mineralization of compounds; iv) sorption of contaminants on the cell surface; v) extracellular sequestration; and vi) absorption into the cell.[134]

Bacterial cell walls are negatively charged under circumneutral pH conditions, and they can passively adsorb positively charged metal ions.[134] For example, *Pseudomonas aeruginosa* can adsorb metal cations due to the presence of negatively charged groups in the membrane such as carboxyl, hydroxyl, sulfonate, amine, sulfhydryl and phosphonate.[137] In fungi, carboxyl, phosphate, amine, amide and alkane groups are involved in metal binding.[135] Some soil bacteria are also able to decrease the toxicity or solubility of some metals and metalloids by changing their oxidation state, such as As or Cr, mostly after being stimulated by the addition of easily available carbon sources. For example, Lee et al.[139] observed that native bacteria from a contaminated mine soil were able to transform dissolved As(III) to As(V), decreasing solubility and toxicity. However, they were unable to identify the species or genera. Some microbial strains can reduce Cr(VI) to Cr(III), which is a more stable and less toxic form. This reduction reaction is catalyzed by chromate reductase activity present in some bacteria such as *Pseudomonas*, *Bacillus* and *Arthrobacter*, related to soluble cytosolic enzymes or insoluble cell membrane enzymes.[135,140] The biotransformation of Cr(VI) is considered as a detoxification mechanism because the trivalent form is more stable and less toxic than Cr(VI). Some recent studies reported extracellular chromate reductase activity present in the supernatant of microbial cultures.[140,184] In bacteria (including *Pseudomonas, Bacillus* and *Arthrobacter*), the enzymatic Cr(VI) reduction could be related to soluble cytosolic proteins or insoluble cell membrane enzymes.[185] Indeed, different chromate reductases (ChrR, YieF, NemA, and LpDH) were found in the cytoplasmic fraction (soluble) or bound to the membrane catalyzing the reaction under aerobic or anaerobic (or sometimes both) conditions.

In recent years, great interest has been shown in the use of microbially induced carbonate precipitation (MICP) for the reclamation of soils affected by metals. The release of CO_3^{2-} and NH_4^+ ions by ureolytic bacteria (with increasing pH) in the presence of metal cations induces carbonate precipitation. Urea can be degraded as follows:

$$2(NH_2)2CO + 2H_2O \rightarrow CO_3^{2-} + 2NH_3$$

Kang et al.[136] proved that microbial hydrolysis of urea results in fast production of carbonate ions, followed by precipitation of $PbCO_3$ without any calcium source. The identified soil ureolytic bacteria in this study were *Sphingobacterium* sp., *Enterobacter cloacae* and *Lysinibacillus sphaericus*. Kang et al.[141] isolated four bacterial strains from a metalliferous mine soil (*Viridibacillus arenosi* B-21, *Sporosarcina soli* B-22, *E. cloacae* KJ-46, and *E. cloacae* KJ-47) with effective MICP. The use of the mixture of the four bacterial strains reduced the availability of Pb, Cd, and Cu by calcite production. *Pararhodobacter sp.* has also proved to be efficient at reducing Pb availability in mine soils by coprecipitation of $CaCO_3$ and Pb.[142]

Many microorganisms sequester metals outside the cells to avoid their toxic effects. The release of extracellular exopolysaccharides composed of polysaccharides, proteins, nucleic acids, humic substances, lipids and other low molecular weight constituents have the ability to sequester metals such as Pb, Cd, or Cu, reducing their availability.[143,144] Mine soils contaminated with Hg are logistically difficult to reclaim and with high costs. It has been proved that the addition of *Pseudomonas veronii* encapsulated with a biopolymer and immobilized onto a natural zeolite substrate can uptake Hg from the soil and promote its volatilization by passive diffusion from the cell as mercury ions.[145] These researchers recorded Hg emissions 10^4 times above the background levels, and demonstrated that *P. veronii* cells could be stored for at least four months without loss of functionality. Some microbial species synthetize low molecular weight proteins rich in cysteine called metallothioneins, which play a critical role in the immobilization of metals within the cell to counteract toxic effects. Some bacteria such as *Pseudomonas fluorescens, P. aeruginosa, Bacillus megaterium, Bacillus cereus* or *Delftia sp.* resist soil metal toxicity by internal immobilization led by metallothionein protein.[131,146–149]

Despite the promising results for the use of biotechnologies for mine soil reclamation, microorganisms hardly develop in contaminated mine soils, with a very low population and activity.[150] Thus, until more effective strategies based on microorganisms for soil reclamation are developed, field-scale reclamation is ultimately based on the use of soil microorganisms that are associated with plants (phytotechnologies). The symbiosis plant-microorganisms enhances the ability of both types of organisms to survive and grow in extreme environments, contributing to the reclamation of the soils. The tissues and surfaces of a plant hosting microbial communities can be classified into rhizosphere (interphase between roots and soil), phyllosphere (aerial surface), and endosphere (internal plant tissues which host endophyte microbes).

4.3 Phytotechnologies

The phytotechnologies have gradually attracted more attention for reclaiming mine soils, since they are best suited for the remediation of diffusely polluted areas through the use of plants. This strategy has been proved to be technically and economically feasible, and can be performed on site, thus reducing exposure risks for clean-up personnel or secondary contamination in transport.[151] Thus, phytotechnology or phytoremediation can be defined as an efficient, inexpensive, and environmentally- and eco-friendly reclamation technology using vegetation to remove and/or detoxify soils.[152] The efficiency of phytotechnologies for the reclamation of mine soils has been demonstrated in recent years,[153–155] thus, there is now an increasing interest for screening new suitable species growing naturally in contaminated soils. The plants used in phytoremediation present constitutive mechanisms that confer them tolerance to pollutants, with the ability of accumulating high concentration of pollutants in their rhizosphere and tissues.

Three major phytotechnolgies can be defined in terms of plant physiology: phytoextraction, phytovolatilization and phytostabilization.[151,156] Phytoextraction involves the use of plants which are capable of taking up high quantities of pollutants from the soil and transferring them into the aerial tissues, where they are stored with no toxic effect on the plant. The above-ground biomass enriched in pollutants is then disposed of as hazardous waste or incinerated for metal recovery through the ashes. Phytovolatilization involves plants being able to take up pollutants from the soil and release them into the atmosphere through stomata by transpiration. Phytostabilization involves the use of plants that foster the immobilization of contaminants (mostly metals and metalloids) in the rhizosphere or in the roots, hindering the translocation of contaminants to the above-ground tissues. Immobilization can take place by promotion of precipitation in the soil, adsorption to root surface, or accumulation in the roots. This technology enables the growth of vegetation that fosters soil detoxification with no risk of contaminants being transferred to the food chain, since no contaminant is accumulated in the shoots. Furthermore, the vegetation cover created in the area stabilizes the soil from being eroded by air and runoff, and since the toxicity of contaminants is decreased, other plant species can colonize the area, thereby ensuring the long-term reclamation of the soil by creating a self-sustaining vegetation cap.[155]

In many cases, soil conditions hinder the growth and development of vegetation, and therefore, the efficiency of phytotechnologies. To overcome this situation, the use of soil ameliorants is essential. Thus, the success of this phytotechnology is controlled by selecting the right plant species to recover a soil, as well as the careful selection of amendments (organic matter, chelating agents, lime, etc.) that improve soil properties and encourage the survival and growth of plants.[153] The proper selection of amendments also contributes to modifying the mobility of the metals and metalloids, depending on the reclamation objective: to solubilize them for phytoextraction or to precipitate them for phytostabilization.[157] Any amendment described in the section on the use of physicochemical techniques to reclaim mine soils could be used to ameliorate soil conditions so that plants can grow, contributing to shifts in soil pH, improvement of soil structure, increase of soil nutrients, etc. The combined use of phytotechnologies and soil ameliorants is called aided phytoremediation: if we aid the plants to grow, they will aid us to clean up the soil.

To succeed with phytoextraction, the selection of plant species is critical. To guarantee that the process is efficient and feasible, the use of hyperaccumulators should be encouraged. Metal hyperaccumulators are plants that exhibit higher levels of metal or metalloid absorption in their tissues and with no symptoms of phytoxicity, and also show high biomass production, so that the process of decontamination by plant uptake is fast. Hyperaccumulators should also have extended roots systems to explore large soil volumes, high translocation factors to the above-ground tissues and easily respond to agricultural management.[158] *Brassicaceae* species are commonly known to have exceptionally high metal-accumulating capacities.[151] The availability of pollutants, mostly metals, for plant uptake is conditioned by soil factors, such as pH, redox potential, soil adsorption capacity, texture and quantity and quality of root exudates and microbial activity.[157]

Efficiency of phytoextraction can be improved by advanced agronomic practices such as the implementation of genetic engineering to enhance the metal tolerance, shoot translocation, accumulation, and sequestration. Effective practices to improve the absorption rate of metals by plants involve adding soil ameliorants to decrease the pH, so that metals are mobilized, and to add chelating compounds that favor plant uptake.[157,159] The application of bioaugmentation (addition of microorganisms to speed up the rate of decontamination) may facilitate the phytoextraction efficiency.[160] Phytovolatilization is a technique that has not been implemented as widely as phytoextraction and phytostabilization, since it relies on the ability of certain species to absorb and volatilize some metals and metalloids, and not many species show this trait. This phytotechnology is mostly used for Hg, Se and As,[123,161] which are taken up by the roots, transformed into non-toxic species, and then released during the transpiration process of the plant physiology. The plant species most frequently studied for this purpose have been *Astragalus sp., Stanleya sp., Caulanthus sp., Chara sp., Arabidopsis thaliana, Nicotiana tabacum* or *Brassica sp.*[117,123,161,162] The addition of pH modifiers and chelating agents that solubilize metals and metalloids speeds up the phytovolatilization process by more efficient root uptake.[117] Nonetheless, the practical implementation of phytovolatilization has been questioned and criticized because once the metal or metalloid is volatilized, there is no longer control over its fate, with the hazard of provoking pollution in other areas.[163] In this sense, however, Meagher[164] suggested that once in the atmosphere, volatile metals and metalloids are diluted and dispersed with a very low environmental hazard.

With phytovolatilization and phytoextraction strategies, there is a risk of transfer of pollutants to the food chain, so a thorough control of the area is essential to avoid hazards to the ecosystems. Once the soil has been decontaminated, the area should be revegetated with native vegetation or given another land use according to the objectives of the reclamation. Contrarily, phytostabilization can contribute to creating a long-term self-sustaining ecosystem by a reduction in soil toxicity and an improvement in soil quality and biodiversity (Fig. 2). Among vegetation families, those species belonging to the Gramineae family have shown the best attributes for use as phytostabilization candidates, due to their high potential to accumulate large amounts of metals and metalloids in the roots, with a very low translocation ratio and active immobilization of metals and metalloids in the rhizosphere.[138,155,165,166] Fig. 3 shows an image of an experiment carried out in a tailings pond in SE Spain affected by a high content of As, Pb, and

Fig. 3 Image of an experiment carried out in a tailings pond in SE Spain affected by a high content of As, Pb, and Zn, with initially null vegetation (A), and totally vegetated three years after phyostabilization (B).

Zn, with initially null vegetation owing to extreme acidity, high solubility of metals and metalloids, and low organic matter and nutrient contents.[167] Those authors used marble waste as soil ameliorants as a source of carbonates to neutralize acidity, immobilize metals and develop soil structure, and pig slurry and pig manure as a source of organic matter and nutrients for soil development and vegetation establishment. After incorporating the amendments, 10 different plant species with the ability for phytostabilization were introduced. After three years, tailings toxicity decreased, with a 90–99% reduction in metals and metalloids mobility, enabling not only the development of the planted species, but also other native species that were able to colonize the tailings from the surroundings, now that the soil conditions were more favorable. *Piptaherum miliaceum* and *Lygeum spartum* were the species with the highest Zn, Pb, and As concentrations in the roots, with no translocation to the aerial tissues, presenting themselves as good candidates for phytostabilization of tailings ponds. Most soil ameliorants used for aiding phytostabilization have been based on compounds that increase soil pH (such as calcium carbonate, lime, fly ash, etc.) to decrease metal mobility, and organic matter to provide nutrients and foster soil structure (manure, compost, sludge, biochar, etc.).[101,168,169]

5. Good management of mining wastes to avoid soil degradation

There are large agricultural and natural areas close to mining areas that may suffer the environmental impacts caused by mining wastes; therefore comprehensive efforts are necessary to ensure that the risk of contamination of fertile soil in agricultural, horticultural and natural lands is reduced to the minimum. The problem of loss of soil fertility of the aforementioned lands can be addressed by planning and executing an appropriate drainage system in and around the mining or quarrying area in such a way that the suspended sediments in the runoff water are arrested by devices such as gully plugs, check dams, silt settling tanks, etc.

It is important that the disposal of mining waste on surface soil is adequate to minimize its erosion and decrease wind and water transport of particles contaminated by metals and metalloids. Erosion of soil due to water is classified into three types,[170] namely sheet erosion, rill erosion and gully erosion. It is well established that surface erosion can occur as a result of flowing water, and slope angle is one of the major factors affecting water erosion. Therefore, steeper slopes lead to greater erosion as a consequence of more splash, higher runoff and faster velocities. Williams (1996)[171] argued that soil erosion in mining is a function of: (a) method of rehabilitation; (b) slope angle; (c) slope length; (d) slope profile, whether convex, uniform, or concave; (e) physical and chemical nature of the spoil; and (f) presence and extent of surface rock. The effects of vegetation on soil erosion have been studied by Nyakatawa et al.[172] and Fadda et al.[173] and all those studies have equivocally observed that there is a great potential for reduction in runoff and erosion due to large vegetation cover. However, they noted that the effect varies depending on the physical arrangements of the cover and the size of cover elements. As suggested by Morgan,[174] it is imperative to provide erosion-resistant covers, such as vegetation, waste rock or surface crusting to reduce the velocity across the disturbed areas. Furthermore, the authors observed that provision of diversion terraces, contour bunds and waterways could also help in reducing soil erosion in addition to covering the soil surface by the application of materials such as mulch, other natural materials or synthetic materials, and ripraps.

The genesis of acid mine drainage is one of the main problems associated with the mining of metal sulfides that may affect the soils near the mining areas, and the main physical, chemical and biological factors that have been studied in recent years,[175–177] and models have been developed to evaluate the acid potential of the waste, that is, to predict the time that acidity can remain in the medium. Moreover, many of these investigations have focused on studying the most appropriate management of this acid mine drainage to reduce soil degradation. Possible measures to prevent or significantly reduce acid rock drainage include:[178]

- Flooding of old underground mine workings to cut off the oxygen supply necessary for the sustained generation of acidic waters.
- Sealing exposed surfaces in underground workings with a coating of material that is non-reactive or impermeable to inhibit the oxidation process.
- Backfilling mine workings with reactive materials that can neutralize and treat waters that pass through them.
- Adding chemicals to the water in flooded surface and underground mine workings that can inhibit acid-generating chemical reactions and precipitate coatings that will seal off groundwater migration routes.
- Isolating contaminated waters at depth by stratification, allowing a viable habitat to develop near the surface in the water that fills large open pits.

Adequate measures should also be taken to cover the waste dumps with vegetation in order to reduce the chances of erosion and leaching. Currently, phytoremediation techniques are widespread in the reclamation of mining areas around the world.

Other possible measures that may be applied in the mine areas to restrict dust pollution so that its spread to the nearby agricultural and horticultural lands is minimized could be:

- Assessment of environmental risks in connection with the deposition of soil particles;
- Characterization of physical-chemical properties of mining wastes;
- Development of locally adapted measures for landfills and old polluted mine tailings, with a focus on long-term sustainable solutions;

- The use of model tests in the lab to describe and predict the behavior and dispersion of contaminants;
- Review of waste reuse in structures such as fills, roads and barriers;
- Mapping of natural geological barriers and both natural materials and residual products for the top sealing of landfills.

6. Conclusions

Metallic and non-metallic mine activity provides materials involved in many other industrial activities such as building and road construction, automobile industry, computer and satellite components or in power generation. Therefore, mining supposes a great proportion of the GDP in those countries where the mining activity is developed. However, during the development and exploitation stages several environmental impacts occur linked to the creation of new roads and buildings, soil removal and waste generation. Hence, mine activities cause visual and ecological impacts, changing the landscape and breaking ecosystems.

Major activities related to changes in landscapes are the soil removal and its consequent loss of vegetation cover. In metallic mining, dumping of mine spoils in surrounding mining areas further complicate plant establishment due to the physicochemical conditions of the residues. The pH falls to very low values (pH2−3), causing a considerable release of Fe, Al, and toxic trace metals while changes in the redox state of an environment cause strong changes in the behavior and mobility of metals such as Pb, Cd, and Zn. Moreover, mining activity also results in structural and functional changes in mine soils which can be clustered into three types of degradation: physical, chemical and biological. Physical degradation is mainly related to the loss of soil structure, while chemical and biological degradation is connected to the lack of nutrients and the growth of micro and mesofauna in mining environments. Reclamation activities in mine tailings are mandatory to avoid phytotoxicity and to supply the nutrients necessary for good plant establishment that promotes a vegetation cover through which the landscape and ecological communities are restored. Therefore, the goal of mine reclamation is not necessarily to restore pre-disturbance conditions, but to reduce hazards to the environment and public health and improve the esthetic value of the area. Most techniques used for soil reclamation worldwide are based on physicochemical techniques, bioremediation (use of microorganisms) or phytotechnologies (use of vegetation associated to the symbiotic microorganisms).

The addition of topsoil, amendments or Technosols (technical soils) are efficient strategies for mine soil reclamation by pH control and the improvement in soil organic matter and nutrients, critical to reducing pollution by toxic elements and to promote the development of a self-sustaining environment. When there is a need for removal of contaminants, other techniques such as electrokinetics, soil washing, or soil flushing can be selected, with the subsequent addition of organic amendments to recover soil fertility.

To reduce the toxicity of toxic mine compounds, microorganisms or vegetation can be used, owing to their ability to decontaminate the areas or reduce the toxicity of the contaminants. The use of biotechnologies implies the use of tolerant microorganisms to reclaim mine soils, taking advantage of the biogeochemical cycles in which, they are specifically involved. However, the limitation for this strategy is the very limited information available on the diversity of microorganisms living in mine soils.

Phytotechnologies are used for the in-situ remediation of diffusely polluted areas through the use of plants. Three major phytotechnologies can be defined in terms of plant physiology: phytoextraction (accumulation of pollutants in above-ground biomass which are later disposed); phytovolatilization (uptake of pollutants by plants and posterior release into the atmosphere by transpiration); and phytostabilization (immobilization of pollutants in the rhizosphere and/or plant roots, decreasing environmental hazards). In many cases, soil conditions hinder the growth and development of vegetation, and thus, the efficiency of phytotechnologies. To overcome this situation, the use of soil ameliorants is essential.

Actions focused on reducing the effects of climate agents through the wastes disposed in tailings ponds are important to prevent soil degradation in the surrounding areas as well in the tailing ponds themselves. Thus, the provision of diversion terraces, contour bunds and waterways could also help in reducing soil erosion in addition to covering the soil surface by the application of materials such as mulch, other natural materials or synthetic materials, and ripraps. Other activities should additionally be focused on the reduction of conditions that promote the acid drain formation or the wind and water spreading fine particles containing high amounts of metals and metalloids, which are potentially harmful to agricultural lands and for human health.

References

1. Darko E. *The existence of the Chirano Gold Mine limited and its effect on the livelihood of the Chirano concession area*. Kumasi College of Engineering: Kwame NKrumah University of science and technology; 2012.
2. Madigan RT. *Of minerals and man*. Australia: Austalasian lust. Mag and Wetportville; 1981. p. 138.
3. Walser G. *Economic impact of world mining*, **1 1**. World Bank Group Mining Department; 2010. p. 86–8.
4. SERI and Wu-global material flow database, Sustainable Europe Research Institute (SERI) & Vienna University of Economics and Business (WU) 2014.
5. Sandell E, Tigabu M, Chileshe M, Syampungani S, Odén PC. Progresses in restoration of post-mining landscape in Africa. *J For Res* 2019;**30**:381–96.
6. Hartman LH, Mutmansky JM. *Introduction to mining engineering*. 2nd ed. John Willey & Sons; 2002. p. 633. http://www.edumine.com/xtoolkit/xmlwords/xmlwords.htm.
7. Haldar SK, Tišljar J. *Chapter 8 - mineral deposits: host rocks and origin introduction to mineralogy and petrology*. 2014. p. 261–79.
8. Haldar SK. *Introduction to mineralogy and petrology*. Oxford: Elsevier; 2014.
9. Deshmukh DJ. *Elements of mining technology*, vol. 2. Vidyaseva Prakashan; 2016. p. 456.
10. Deshmukh DJ. *Elements of mining technology*, vol. 3. Denett & Company; 2016. p. 472.
11. Haldar SK. *Chapter 11 - elements of mining*. Mineral Exploration; 2018. p. 193–222. https://doi.org/10.1016/B978-0-12-814022-2.00012-5.
12. *Gold fields: mineral resource and mineral reserve overview*. 2011. Africa.
13. Environmental Law Alliance Worldwide (ELAW). *Guidebook for evaluating mining project EIAs*. USA. 2010. p. 122.
14. Essalhi A, Essalhi M, Toummite A. Environmental impact of mining exploitation: a case study of some mines of barite in the eastern Anti-Atlas of Morocco. *J Environ Prot* 2016;**7**:1473–82.
15. Zanbak C, Heap Leaching Technique in Mining. *Euromines – the European Association of mining industries, metal ores & industrial minerals*. 2012.
16. IFC/World Bank. *Environmental, health and safety guidelines for mining*. 2007.
17. EIA Technical Review Guidelines: non-Metal and Metal Mining. *CAFTA-DR and US country experts supported by USAID-environment and labor excellence-CCAD-USEPA program*. 2011.
18. Koryagin M, Voronov A. Improving the organization of the shovel-truck systems in open-pit coal mines. *Trans Prob* 2017;**12**(2):113–22.
19. Hustrulid WA, Bullock RL. *Underground mining methods engineering fundamentals and international case studies*. USA: Society for Mining, Metallurgy, and Exploration, Inc., (SME); 2001. p. 718.
20. Briggs DF. *Recovery of copper by solution mining methods*. Arizona Geological Survey Contributed Report CR-115-A; 2015. p. 10.
21. Management BRGM. *Of mining, quarrying and ore-processing waste in the European Union*. 2011. p. 79.
22. Garbarino E, Orveillon G, Saveyn Hans GM, Barthe P, Eder P. *Best available techniques (BAT) reference document for the management of waste from extractive industries*. European Commission; 2018.
23. Schoenberger E. Environmentally sustainable mining: the case of tailings storage facilities. *Resour Policy* 2016;**49**:119–28.
24. NSW. *Design report for the Blackwood pit taoiling storage facility extensión*. Golder Associates; 2017.
25. Seervi V, Yadav HL, Srivastav SK, Jamal A. Overview of active and passive systems for treating acid mine drainage. *Int Adv Res J Sci Eng Technol* 2017;**4**(5):131–7.

26. RoyChowdhury A, Sarkar D, Datta R. Remediation of acid mine drainage-impacted water. *Curr Pollut Rep* 2015;**1**:131–41.
27. Pearse MJ. An overview of the use of chemical reagents in mineral processing. *Miner Eng* 2005;**18**(2):139–49.
28. Mutsvanga T, Mapira J, Ngaza N. Gold ore waste management challenges at Renco Mine, Zimbabwe. *Eur J Social Sci Stud* 2018;**3**(3):18–34.
29. Sklenička P, Kašparová I. Restoration of visual values in a post-mining landscape. *J Landscape Stud* 2008;**1**:1–10.
30. Qian D, Yan C, Xing Z, Xiu L. Monitoring coal mine changes and their impact on landscape patterns in an alpine region: a case study of the muli coal mine in the Qinghai-Tibet Plateau. *Environ Monit Assess* 2017;**189**:559.
31. Strunhsacker DW. Chapter 7. Environmental permitting. In: Marcus JJ, editor. *Mining environmental handbook*. Imperial College Press; 1997. p. 313.
32. Monjezi M, Shahriar K, Dehghani H, Samimi Namin F. Environmental impact assessment of open pit mining in Iran. *Environ Geol* 2009;**58**(1):205–16.
33. Singh AN, Raghubanshi AS, Singh JS. Comparative performance and restoration potential of two Albizia species planted on mine spoil in a dry tropical region, India. *Ecol Eng* 2004;**22**:123–40.
34. Altun OA, Yilmaz I, Yildirim M. A short review on the surficial impacts of underground mining. *Sci Res Essays* 2010;**5**(21):3206–12.
35. Tordoff GM, Baker AJM, Willis AJ. Current approaches to the revegetation and reclamation of metalliferous mine wastes. *Chemosphere* 2000;**41**:219–28.
36. Wong MH. Ecological restoration of mine degraded soils, with emphasis on metal contaminated soils. *Chemosphere* 2003;**50**:775–80.
37. Hooda PS. *Trace elements in soils*. Blackwell Publishing Ltd.; 2010, ISBN 978-1- 405-16037-7.
38. Kabata-Pendias A, Pendias H. *Trace elements in soils and plants*. 2nd ed. CRC Press; 1992.
39. Kabata-Pendias A. Soil-plant transfer of trace elements—an environmental issue. *Geoderma* 2004;**122**:143–9.
40. Gerasimova MI, Stroganova MN, Mozharova NV, Prokof'eva TV. *Anthropogenic soils: genesis, geography, and rehabilitation*. Smolensk: Oikumena; 2003.
41. Anju M, Banerjee DK. Multivariate statistical analysis of heavy metals in soils of a Pb–Zn mining area, India. *Environ Monit Assess* 2012;**184**:4191–206.
42. Chakraborty S, Man T, Paulette L, Deb S, Li B, Weindorf DC, Frazier M. Rapid assessment of smelter/mining soil contamination via portable X-ray fluorescence spectrometry and indicator kriging. *Geoderma* 2017;**306**:108–19.
43. Ding Q, Cheng G, Wang Y, Zhuang D. Effects of natural factors on the spatial distribution of heavy metals in soils surrounding mining regions. *Sci Total Environ* 2017;**578**:577–85.
44. Lehmann A. Technosols and other proposals on urban soils for the WRB (world reference base for soil resources). *Int Agrophys* 2006;**20**:129–34.
45. Zipper CE, Burger JA, Barton CD, Skousen JG. Rebuilding soils on mined land for native forests in Appalachia. *Soil Sci Soc Am J* 2013;**77**(2):337–49.
46. Zhang GJ, Bai ZK, Wang JM. Preliminary study on cognition and utilization of mine soils. *China Mining Mag* 2014;**03**:53–6.
47. Ahirwal J, Maiti SK. Development of technosol properties and recovery of carbon stock after 16 years of revegetation on coal mine degraded lands, India. *Catena* 2018;**166**:114–23.
48. Thomas K, Sencindiver J, Skousen J, Gorman J. Soil development on a mountaintop removal mine in southern West Virginia. In: *Proceedings of 2000 annual meeting of the American Society for surface mining and reclamation, Tampa, FL*; 2000. p. 546–56.

49. Liu RQ, Lal R. A laboratory study on amending mine soil quality. *Water Air Soil Pollut* 2013;**224**(9).
50. Rowland SM, Prescott CE, Grayston SJ, Quideau SA, Bradfield GE. Recreating a functioning forest soil in reclaimed oil sands in northern Alberta: an approach for measuring success in ecological restoration. *J Environ Qual* 2009;**38**(4):1580–90.
51. Vanbreemen N. Soils as biotic constructs favoring net primary productivity. *Geoderma* 1993;**57**(3):183–211.
52. Jenny H. *Factors of soil formation*. New York, NY, USA: McGraw-Hill Book Company; 1941. p. 241.
53. Indorante SJ, Jansen IJ, Boast CW. Surface mining and reclamation: initial changes in soil character. *J Soil Water Conserv* 1981;**36**(6):347–51.
54. Hartmann M, Niklaus PA, Zimmermann S, Schmutz S, Kremer J, Abarenkov K, Luscher P, Widmer F, Frey B. Resistance and resilience of the forest soil microbiome to logging-associated compaction. *ISME J* 2014;**8**(1):226–44.
55. Mensah AK. Role of revegetation in restoring fertility of degraded mined soils in Ghana: a review. *Int J Biodivers Conserv* 2015;**7**(2):57–80.
56. Shrestha RK, Lal R. Changes in physical and chemical properties of soil after surface mining and reclamation. *Geoderma* 2011;**161**(3):168–76.
57. Ramani RV. Surface mining technology: progress and prospects. *Procedia Eng* 2012;**46**: 9–21.
58. Šourková M, Frouz J, Šantrůčková H. Accumulation of carbon, nitrogen and phosphorus during soil formation on alder spoil heaps after brown-coal mining, near Sokolov (Czech Republic). *Geoderma* 2005;**124**(1):203–14.
59. Mueller L, Schindler U, Mirschel W, GrahamShepherd T, Ball BC, Helming K, Rogasik J, Eulenstein F, Wiggering H. Assessing the productivity function of soils. A review. *Agronomy Sustain Dev* 2010;**30**(3):601–14.
60. Panayiotopoulos KP, Papadopoulou CP, Hatjiioannidou A. Compaction and penetration resistance of an Alfisol and Entisol and their influence on root growth of maize seedlings. *Soil Tillage Res* 1994;**31**(4):323–37.
61. Shrestha RK, Lal R. Land use impacts on physical properties of 28 years old reclaimed mine soils in Ohio. *Plant Soil* 2008;**306**(1):249–60.
62. Lipiec J, Hatano R. Quantification of compaction effects on soil physical properties and crop growth. *Geoderma* 2003;**116**(1):107–36.
63. Khademi H, Abbaspour A, Martínez-Martínez S, Gabarrón M, Faz A, Acosta JA. Provenance and environmental risk of windblown materials from mine tailing ponds, Murcia, Spain. *Environ Pollut* 2018;**241**:432–40.
64. Zhen Q, Ma WM, Li MM, He HH, Zhang XC, Wang Y. Effects of vegetation and physicochemical properties on solute transport in reclaimed soil at an opencast coal mine site on the Loess Plateau, China. *Catena* 2015;**133**:403–11.
65. Lal R. Restoring soil quality to mitigate soil degradation. *Sustainability* 2015;**7**:5875–95.
66. Wei LL, Wang K, Noguera DR, Jiang JQ, Oyserman B, Zhao NB. Trans-formation and speciation of typical heavy metals in soil aquifer treatment system during long time recharging with secondary effluent: depth distribution and combination. *Chemosphere* 2016;**165**:100–9.
67. Larocque ACL, Rasmussen PE. An overview of trace metals in the environment, from mobilization to remediation. *Environ Geol* 1998;**33**(2):85–91.
68. Gisore R, Matina Z. *Sustainable mining in Africa: standards as essential catalysts*. Kenya: ARSO Central Secretariat Nairobi; 2015. p. 55.
69. Martínez-Martínez S, Acosta JA, Cano AF, Carmona DM, Zornoza R. Assessment of the lead and zinc contents in natural soils and tailing ponds from the Cartagena-La Unión mining district, SE Spain. *J Geochem Explor* 2013;**124**:166–75.

70. Machín J, Navas A. Soil pH changes induced by contamination by magnesium oxides dust. *Land Degrad Dev* 2000;**11**:37–50.
71. Zhou QX. *Health soil science: soil health quality and safety of agricultural products.* Science Press Beijing; 2005.
72. Amegbey N. *Lecture notes on environmental engineering in mining.* Tarkwa, Ghana: University of Science and Technology, School of Mines; 2010. p. 1–34. 117-118; 120-122; 124; 127-128; 135; 139. Raft National Mining Policy of Ghana.
73. Assel PG. *Evaluating the usefulness of* Acacia auriculiformis *in ameliorating surface mine degraded lands.* B.Sc. Dissertation, 1-24. Kumasi, Ghana: Department of Agroforestry, Kwame Nkrumah University of Science and Technology; 2006.
74. Hilson G. A contextual review of the Ghanaian small-scale mining industry. *Mining, Minerals Sustain Dev* 2001;**76**.
75. Sheoran V, Sheoran AS, Poonia P. Soil reclamation of abandoned mine land by revegetation: a review. *Int J Soil Sediment Water* 2010;**3**(2):13.
76. Doran JW, Parkin TB. Defining and assessing soil quality. In: Doran JW, Coleman DC, Bezdicek DF, Stewart BA, editors. *Defining soil quality for a sustainable environment.* Madison: Soil Science Society of America and American Society of Agronomy; 1994. p. 3–21.
77. Frouz J, Kalcik J, Velichova V. Factors causing spatial heterogeneity in soil properties, plant cover, and soil fauna in a non-reclaimed post-mining site. *Ecol Eng* 2011;**37**(11): 1910–3.
78. Feng Y, Wang J, Bai Z, Reading L. Effects of surface coal mining and land reclamation on soil properties: a review. *Earth Sci Rev* 2019;**191**:12–25.
79. Maiti SK. Biodiversity erosion and conservation in Ecorestored site. Ecorestoration of the coalmine degraded lands. *Springerlink* 2012:187–99.
80. Beare MH. Fungal and bacterial pathways of organic matter decomposition and nitrogen mineralization in arable soils. *J Opt Soc Am* 1997;**55**(5):574–5.
81. arney FJ, Angers DA. The role of organic amendments in soil reclamation: a review. *Can J Soil Sci* 2012;**92**(1):19–38.
82. Kaschuk G, Alberton O, Hungria M. Three decades of soil microbial biomass studies in Brazilian ecosystems: Lessons learned about soil quality and indications for improving sustainability. *Soil Biol Biochem* 2010;**42**(1):1–13.
83. Filcheva E, Noustorova M, Gentcheva-Kostadinova S, Haigh MJ. Organic accumulation and microbial action insurface coal-minespoils, Pernik, Bulgaria. *Ecol Eng* 2000; **15**(1):1–15.
84. Frouz J, Elhottova D, Kuraz V, Sourkova M. Effects of soil macrofauna on other soil biota and soil formation in reclaimed and unreclaimed post mining sites: results of a field microcosm experiment. *Appl Soil Ecol* 2006;**33**(3):308–20.
85. Helingerova M, Frouz J, Santruckova H. Microbial activity in reclaimed and unreclaimed post-mining sites near Sokolov (Czech Republic). *Ecol Eng* 2010;**36**(6): 768–76.
86. Claassens S, vanRensburg PJ, Liebenberg D, vanRensburg L. A comparison of microbial community function and structure in rehabilitated asbestos and coal discard sites. *Water Air Soil Pollution* 2012;**223**(3):1091–100.
87. Castaldi S, Rutigliano FA, Virzo de Santo A. Suitability of soil microbial parameters as indicators of heavy metal pollution. *Water Air Soil Pollut* 2004;**158**:21–35.
88. Angelovičová L, Fazekašová D. Contamination of the soil and water environment by heavy metals in the former mining area of Rudňany (Slovakia). *Soil Water Res* 2014;**9**: 18–24.
89. Alkorta I, Aizpurua A, Riga P, Albizu I, Amézaga I, Garbisu C. Soil enzyme activities as biological indicators of soil health. *Rev Environ Health* 2003;**18**:65–73.

90. Hu XF, Jiang Y, Shu Y, Hu X, Liu L, Lua F. Effects of mining wastewater discharges on heavy metal pollution and soil enzyme activities on the paddy fields. *J Geochem Explor* 2014;**147**:139–50.
91. Mikanova O. Effects of heavy metals on some biological parameters. *J Geochem Explor* 2006;**88**:220–3.
92. D'Ascoli R, Rao MA, Adamo P, Renella G, Landi L, Rutigliano FA, Terribile F, Gianfreda L. Impact of river overflowing on trace element contamination of volcanic soils in south Italy: part II. Soil biological and biochemical properties in relation to trace element speciation. *Environ Pollut* 2006;**144**:317–26.
93. Bhattacharya P, Tripathy S, Chakrabarti K, Chakraborty A, Banik P. Fractionation and bioavailability of metals and their impact on microbial properties in sewage irrigated soil. *Chemosphere* 2008;**72**:543–50.
94. Pattnaik BK, Equeenuddin SM. Potentially toxic metal contamination and enzyme activities in soil around chromite mines at Sukinda Ultramafic complex. *India J Geochem Explor* 2016;**168**:127–36.
95. Bradshaw AD. *Underlying principles of restoration*. 1996.
96. Swab RM, Lorenz N, Byrd S, Dick R. Native vegetation in reclamation: improving habitat and ecosystem function through using prairie species in mine land reclamation. *Ecol Eng* 2017;**108**:525–36. https://doi.org/10.1016/J.ECOLENG.2017.05.012.
97. Fernández-Calviño D, Cutillas-Barreiro L, Nóvoa-Muñoz JC, Díaz-Raviña M, Fernández-Sanjurjo MJ, Álvarez-Rodríguez E, Núñez-Delgado A, Arias-Estévez M, Rousk J. Using pine bark and mussel shell amendments to reclaim microbial functions in a Cu polluted acid mine soil. *Appl Soil Ecol* 2018;**127**:102–11. https://doi.org/10.1016/J.APSOIL.2018.03.010.
98. Joniec J. Indicators of microbial activity in the assessment of soil condition subjected to several years of reclamation. *Ecol Indic* 2019;**98**:686–93. https://doi.org/10.1016/J.ECOLIND.2018.11.055.
99. Asensio V, Vega FA, Covelo EF. Effect of soil reclamation process on soil C fractions. *Chemosphere* 2014;**95**:511–8. https://doi.org/10.1016/J.CHEMOSPHERE.2013.09.108.
100. Kabas S, Faz A, Acosta JA, Arocena JM, Zornoza R, Martínez-Martínez S, Carmona DM. Marble wastes and pig slurry improve the environmental and plant-relevant properties of mine tailings. *Environ Geochem Health* 2014;**36**. https://doi.org/10.1007/s10653-013-9517-z.
101. Madejón P, Domínguez MT, Madejón E, Cabrera F, Marañón T, Murillo JM. Soil-plant relationships and contamination by trace elements: a review of twenty years of experimentation and monitoring after the Aznalcóllar (SW Spain) mine accident. *Sci Total Environ* 2018;**625**:50–63. https://doi.org/10.1016/J.SCITOTENV.2017.12.277.
102. Moreno-Barriga F, Díaz V, Acosta JA, Muño MÁ, Á F, Zornoza R. Organic matter dynamics, soil aggregation and microbial biomass and activity in technosols created with metalliferous mine residues, biochar and marble waste. *Geoderma* 2017;**301**. https://doi.org/10.1016/j.geoderma.2017.04.017.
103. Novak JM, Ippolito JA, Ducey TF, Watts DW, Spokas KA, Trippe KM, Sigua GC, Johnson MG. Remediation of an acidic mine spoil: miscanthus biochar and lime amendment affects metal availability, plant growth, and soil enzyme activity. *Chemosphere* 2018;**205**:709–18. https://doi.org/10.1016/J.CHEMOSPHERE.2018.04.107.
104. Antonelli PM, Fraser LH, Gardner WC, Broersma K, Karakatsoulis J, Phillips ME. Long term carbon sequestration potential of biosolids-amended copper and molybdenum mine tailings following mine site reclamation. *Ecol Eng* 2018;**117**:38–49. https://doi.org/10.1016/J.ECOLENG.2018.04.001.
105. Cheng CM, Amaya M, Butalia T, Baker R, Walker HW, Massey-Norton J, Wolfe W. Short-term influence of coal mine reclamation using coal combustion residues on

groundwater quality. *Sci Total Environ* 2016;**571**:834–54. https://doi.org/10.1016/J.SCITOTENV.2016.07.061.

106. Moreno-Barriga F, Díaz V, Acosta JA, Muñoz MÁ, Á F, Zornoza R. Creation of technosols to decrease metal availability in pyritic tailings with addition of biochar and marble waste. *L Degrad Dev* 2017;**28**. https://doi.org/10.1002/ldr.2714.
107. Asensio V, Flórido FG, Ruiz F, Perlatti F, Otero XL, Oliveira DP, Ferreira TO. The potential of a technosol and tropical native trees for reclamation of copper-polluted soils. *Chemosphere* 2019;**220**:892–9. https://doi.org/10.1016/J.CHEMOSPHERE.2018.12.190.
108. Pramudita D, Iskandar I, Indarto A. Nano-Enhanced materials for reclamation of mine spoils. *Bio-Geotechnologies Mine Site Rehabil* 2018:201–14. https://doi.org/10.1016/B978-0-12-812986-9.00012-9.
109. Zornoza R, Faz A, Carmon DM, Martínez-martínez S, Acosta JA. Plant cover and soil biochemical properties in a mine tailing pond five years after application of marble wastes and organic amendments. *Pedosphere* 2012;**22**. https://doi.org/10.1016/S1002-0160(11)60188-4.
110. Sebastian A, Nangia A, Prasad MNV. Iron oxide nanoparticles and reclamation of mine sites. *Bio-Geotechnologies Mine Site Rehabil* 2018:181–99. https://doi.org/10.1016/B978-0-12-812986-9.00011-7.
111. Zornoza R, Á F, Carmona DM, Acosta JA, Martínez-Martínez S, De Vreng A. Carbon mineralization, microbial activity and metal dynamics in tailing ponds amended with pig slurry and marble waste. *Chemosphere* 2013;**90**. https://doi.org/10.1016/j.chemosphere.2012.10.107.
112. Favas PJC, Martino LE, Prasad MNV. Abandoned mine land reclamation—challenges and opportunities (Holistic approach). *Bio-Geotechnologies Mine Site Rehabil* 2018:3–31. https://doi.org/10.1016/B978-0-12-812986-9.00001-4.
113. Gent DB, Brick RM, Alshawabkeh AN, Larson SL, Fabian G, Granade S. Bench- and field-scale evaluation of chromium and cadmium extraction by electrokinetics. *J Hazard Mater* 2004;**110**:53–62. https://doi.org/10.1016/J.JHAZMAT.2004.02.036.
114. Virkutyte J, Sillanpää M, Latostenmaa P. Electrokinetic soil remediation — critical overview. *Sci Total Environ* 2002;**289**:97–121. https://doi.org/10.1016/S0048-9697(01)01027-0.
115. Ortiz-Soto R, Leal D, Gutierrez C, Aracena A, Rojo A, Hansen HK. Electrokinetic remediation of manganese and zinc in copper mine tailings. *J Hazard Mater* 2019;**365**:905–11. https://doi.org/10.1016/J.JHAZMAT.2018.11.048.
116. Liu L, Li W, Song W, Guo M. Remediation techniques for heavy metal-contaminated soils: principles and applicability. *Sci Total Environ* 2018;**633**:206–19. https://doi.org/10.1016/J.SCITOTENV.2018.03.161.
117. Gong Y, Zhao D, Wang Q. An overview of field-scale studies on remediation of soil contaminated with heavy metals and metalloids: technical progress over the last decade. *Water Res* 2018;**147**:440–60. https://doi.org/10.1016/J.WATRES.2018.10.024.
118. Guardia G, Abalos D, García-Marco S, Quemada M, Alonso-Ayuso M, Cárdenas LM, Dixon ER, Vallejo A. Effect of cover crops on greenhouse gas emissions in an irrigated field under integrated soil fertility management. *Biogeosciences* 2016;**13**:5245–57. https://doi.org/10.5194/bg-13-5245-2016.
119. Pedersen KB, Jensen PE, Ottosen LM, Barlindhaug J. The relative influence of electrokinetic remediation design on the removal of As, Cu, Pb and Sb from shooting range soils. *Eng Geol* 2018;**238**:52–61. https://doi.org/10.1016/J.ENGGEO.2018.03.005.
120. Rosestolato D, Bagatin R, Ferro S. Electrokinetic remediation of soils polluted by heavy metals (mercury in particular). *Chem Eng J* 2015;**264**:16–23. https://doi.org/10.1016/J.CEJ.2014.11.074.

121. Zhou M, Xu J, Zhu S, Wang Y, Gao H. Exchange electrode-electrokinetic remediation of Cr-contaminated soil using solar energy. *Sep Puri Technol* 2018;**190**:297–306. https://doi.org/10.1016/J.SEPPUR.2017.09.006.
122. Zulfiqar W, Iqbal MA, Butt MK. Pb2+ ions mobility perturbation by iron particles during electrokinetic remediation of contaminated soil. *Chemosphere* 2017;**169**:257–61. https://doi.org/10.1016/J.CHEMOSPHERE.2016.11.083.
123. Khalid S, Shahid M, Niazi NK, Murtaza B, Bibi I, Dumat C. A comparison of technologies for remediation of heavy metal contaminated soils. *J Geochemical Explor* 2017;**182**:247–68. https://doi.org/10.1016/J.GEXPLO.2016.11.021.
124. Beiyuan J, Tsang DCW, Valix M, Baek K, Ok YS, Zhang W, Bolan NS, Rinklebe J, Li XD. Combined application of EDDS and EDTA for removal of potentially toxic elements under multiple soil washing schemes. *Chemosphere* 2018;**205**:178–87. https://doi.org/10.1016/J.CHEMOSPHERE.2018.04.081.
125. Hughes DL, Afsar A, Laventine DM, Shaw EJ, Harwood LM, Hodson ME. Metal removal from soil leachates using DTPA-functionalised maghemite nanoparticles, a potential soil washing technology. *Chemosphere* 2018;**209**:480–8. https://doi.org/10.1016/J.CHEMOSPHERE.2018.06.121.
126. Wang G, Zhang S, Zhong Q, Xu X, Li T, Jia Y, Zhang Y, Peijnenburg WJGM, Vijver MG. Effect of soil washing with biodegradable chelators on the toxicity of residual metals and soil biological properties. *Sci Total Environ* 2018;**625**:1021–9. https://doi.org/10.1016/J.SCITOTENV.2018.01.019.
127. Lee YC, Kim EJ, Ko DA, Yang JW. Water-soluble organo-building blocks of aminoclay as a soil-flushing agent for heavy metal contaminated soil. *J Hazard Mater* 2011;**196**:101–8. https://doi.org/10.1016/J.JHAZMAT.2011.08.077.
128. Yan DYS, Lo IMC. Pyrophosphate coupling with chelant-enhanced soil flushing of field contaminated soils for heavy metal extraction. *J Hazard Mater* 2012;**199–200**:51–7. https://doi.org/10.1016/J.JHAZMAT.2011.10.054.
129. Yan X, Liu Q, Wang J, Liao X. A combined process coupling phytoremediation and in situ flushing for removal of arsenic in contaminated soil. *J Environ Sci* 2017;**57**:104–9. https://doi.org/10.1016/J.JES.2016.10.015.
130. Yun SM, Kang CS, Kim J, Kim HS. Evaluation of soil flushing of complex contaminated soil: an experimental and modeling simulation study. *J Hazard Mater* 2015;**287**:429–37. https://doi.org/10.1016/J.JHAZMAT.2015.01.062.
131. Liu Y, Tie B, Li Y, Lei M, Wei X, Liu X, Du H. Inoculation of soil with cadmium-resistant bacterium Delftia sp. B9 reduces cadmium accumulation in rice (Oryza sativa L.) grains. *Ecotoxicol Environ Saf* 2018;**163**:223–9. https://doi.org/10.1016/J.ECOENV.2018.07.081.
132. Thavamani P, Samkumar RA, Satheesh V, Subashchandrabose SR, Ramadass K, Naidu R, Venkateswarlu K, Megharaj M. Microbes from mined sites: harnessing their potential for reclamation of derelict mine sites. *Environ Pollut* 2017;**230**:495–505. https://doi.org/10.1016/J.ENVPOL.2017.06.056.
133. Goldstein A, Lester T, Brown J. Research on the metabolic engineering of the direct oxidation pathway for extraction of phosphate from ore has generated preliminary evidence for PQQ biosynthesis in *Escherichia coli* as well as a possible role for the highly conserved region of quinoprote. *Biochim Biophys Acta* 2003;**1647**:266–71. https://doi.org/10.1016/S1570-9639(03)00067-0.
134. Chon HT, Lee JU, Lee J. Heavy metals contamination of mine soil, their risk assessment, and bioremediation. *Assess Restor Reclam Min Influ Soils* 2017:387–417. https://doi.org/10.1016/B978-0-12-809588-1.00015-3.
135. Fernández PM, Viñarta SC, Bernal AR, Cruz EL, Figueroa LIC. Bioremediation strategies for chromium removal: current research, scale-up approach and

future perspectives. *Chemosphere* 2018;**208**:139–48. https://doi.org/10.1016/J.CHEMOSPHERE.2018.05.166.
136. Kang CH, Oh SJ, Shin Y, Han SH, Nam IH, So JS. Bioremediation of lead by ureolytic bacteria isolated from soil at abandoned metal mines in South Korea. *Ecol Eng* 2015;**74**:402–7. https://doi.org/10.1016/J.ECOLENG.2014.10.009.
137. Kushwaha A, Hans N, Kumar S, Rani R. A critical review on speciation, mobilization and toxicity of lead in soil-microbe-plant system and bioremediation strategies. *Ecotoxicol Environ Saf* 2018;**147**:1035–45. https://doi.org/10.1016/J.ECOENV.2017.09.049.
138. Mahar A, Wang P, Ali A, Awasthi MK, Lahori AH, Wang Q, Li R, Zhang Z. Challenges and opportunities in the phytoremediation of heavy metals contaminated soils: a review. *Ecotoxicol Environ Saf* 2016;**126**:111–21. https://doi.org/10.1016/J.ECOENV.2015.12.023.
139. Lee JU, Lee SW, Kim KW, Yoon CH. The effects of different carbon sources on microbial mediation of arsenic in arsenic-contaminated sediment. *Environ Geochem Health* 2005;**27**:159–68. https://doi.org/10.1007/s10653-005-0133-4.
140. Dey S, Paul AK. Evaluation of chromate reductase activity in the cell-free culture filtrate of Arthrobacter sp. SUK 1201 isolated from chromite mine overburden. *Chemosphere* 2016;**156**:69–75. https://doi.org/10.1016/J.CHEMOSPHERE.2016.04.101.
141. Kang CH, Kwon YJ, So JS. Bioremediation of heavy metals by using bacterial mixtures. *Ecol Eng* 2016;**89**:64–9. https://doi.org/10.1016/J.ECOLENG.2016.01.023.
142. Mwandir W, Nakashima K, Kawasaki S. Bioremediation of lead-contaminated mine waste by Pararhodobacter sp. based on the microbially induced calcium carbonate precipitation technique and its effects on strength of coarse and fine grained sand. *Ecol Eng* 2017;**109**:57–64. https://doi.org/10.1016/J.ECOLENG.2017.09.011.
143. Bramhachari PV, Kavi Kishor PB, Ramadevi R, Kumar R, Rama Rao B, Dubey SK. Isolation and characterization of mocous exopolysaccharide (EPS) produced by Vibrio furnissii strain VB0S3. *J Microbiol Biotechnol* 2007;**17**:44–51.
144. Kushwaha A, Rani R, Kumar S, Thomas T, David AA, Ahmed M. A new insight to adsorption and accumulation of high lead concentration by exopolymer and whole cells of lead-resistant bacterium Acinetobacter junii L. Pb1 isolated from coal mine dump. *Environ Sci Pollut Res* 2017;**24**:10652–61. https://doi.org/10.1007/s11356-017-8752-8.
145. McCarthy D, Edwards GC, Gustin MS, Care A, Miller MB, Sunna A. An innovative approach to bioremediation of mercury contaminated soils from industrial mining operations. *Chemosphere* 2017;**184**:694–9. https://doi.org/10.1016/J.CHEMOSPHERE.2017.06.051.
146. Liu X, Feike T, Shao L, Sun H, Chen S, Zhang X. Effects of different irrigation regimes on soil compaction in a winter wheat-summer maize cropping system in the North China Plain. *Catena* 2016. https://doi.org/10.1016/j.catena.2015.08.014.
147. Murthy S, Bali G, Sarangi S. Effect of lead on metallothionein concentration in lead-resistant bacteria Bacillus cereus isolated from industrial effluent. *AFRICAN J Biotechnol* 2011;**10**:15966–72. https://doi.org/10.5897/AJB11.1645.
148. Naik MM, Pandey A, Dubey SK. *Pseudomonas aeruginosa* strain WI-1 from Mandovi estuary possesses metallothionein to alleviate lead toxicity and promotes plant growth. *Ecotoxicol Environ Saf* 2012;**79**:129–33. https://doi.org/10.1016/J.ECOENV.2011.12.015.
149. Sharma A, Johri B, Sharma A, Glick B. Plant growth-promoting bacterium Pseudomonas sp. strain GRP3 influences iron acquisition in mung bean (Vigna radiata L. Wilzeck). *Soil Biol Biochem* 2003;**35**:887–94. https://doi.org/10.1016/S0038-0717(03)00119-6.

150. Zornoza R, Acosta JA, Faz A, Bååth E. Microbial growth and community structure in acid mine soils after addition of different amendments for soil reclamation. *Geoderma* 2016;**272**. https://doi.org/10.1016/j.geoderma.2016.03.007.
151. Wang L, Ji B, Hu Y, Liu R, Sun W. A review on in situ phytoremediation of mine tailings. *Chemosphere* 2017;**184**:594–600. https://doi.org/10.1016/J.CHEMOSPHERE.2017.06.025.
152. Glick BR. Using soil bacteria to facilitate phytoremediation. *Biotechnol Adv* 2010;**28**: 367–74. https://doi.org/10.1016/J.BIOTECHADV.2010.02.001.
153. Clemente R, Walker DJ, Bernal MP. Uptake of heavy metals and as by Brassica juncea grown in a contaminated soil in Aznalcóllar (Spain): the effect of soil amendments. *Environ Pollut* 2005;**138**:46–58. https://doi.org/10.1016/J.ENVPOL.2005.02.019.
154. Moreno-Barriga F, Á F, Acosta JA, Soriano-Disla M, Martínez-Martínez S, Zornoza R. Use of Piptatherum miliaceum for the phyto management of biochar amended technosols derived from pyritic tailings to enhance soil aggregation and reduce metal(loid) mobility. *Geoderma* 2017;**307**. https://doi.org/10.1016/j.geoderma.2017.07.040.
155. Zornoza R, Acosta JA, Martínez-Martínez S, Faz A, Bååth E. Main factors controlling microbial community structure and function after reclamation of a tailing pond with aided phytostabilization. *Geoderma* 2015:245–6. https://doi.org/10.1016/j.geoderma.2015.01.007.
156. Mendez MO, Maier RM. Phytoremediation of mine tailings in temperate and arid environments. *Rev Environ Sci Bio/Technol* 2008;**7**:47–59. https://doi.org/10.1007/s11157-007-9125-4.
157. Sheoran V, Sheoran AS, Poonia P. Factors affecting phytoextraction: a review. *Pedosphere* 2016;**26**:148–66. https://doi.org/10.1016/S1002-0160(15)60032-7.
158. Song N, Ma Y, Zhao Y, Tang S. Elevated ambient carbon dioxide and Trichoderma inoculum could enhance cadmium uptake of Lolium perenne explained by changes of soil pH, cadmium availability and microbial biomass. *Appl Soil Ecol* 2015;**85**:56–64. https://doi.org/10.1016/J.APSOIL.2014.09.007.
159. Wang J, Xing Y, Li P, Xia J, Liu T, Feng X. Chemically-assisted phytoextraction from metal(loid)s-polluted soil at a typical carlin-type gold mining area in southwest China. *J Clean Prod* 2018;**189**:612–9. https://doi.org/10.1016/J.JCLEPRO.2018.04.082.
160. Zornoza R, Gómez-Garrido M, Martínez-Martínez S, Gómez-López MD, Faz Á. Bioaugmentaton in technosols created in abandoned pyritic tailings can contribute to enhance soil C sequestration and plant colonization. *Sci Total Environ* 2017: 593–4. https://doi.org/10.1016/j.scitotenv.2017.03.154.
161. Favas PJC, Pratas J, Varun M, D'Souza R, Paul MS. Phytoremediation of soils contaminated with metals and metalloids at mining areas: potential of native flora. In: *Environmental risk assessment of soil contamination InTech*; 2014. https://doi.org/10.5772/57469.
162. Ali H, Khan E, Sajad MA. Phytoremediation of heavy metals—concepts and applications. *Chemosphere* 2013;**91**:869–81. https://doi.org/10.1016/J.CHEMOSPHERE.2013.01.075.
163. Ernst WHO. Phytoextraction of mine wastes – options and impossibilities. *Geochemistry* 2005;**65**:29–42. https://doi.org/10.1016/J.CHEMER.2005.06.001.
164. Meagher RB. Phytoremediation of toxic elemental and organic pollutants. *Curr Opin Plant Biol* 2000;**3**:153–62. https://doi.org/10.1016/S1369-5266(99)00054-0.
165. Barbosa B, Fernando AL. Aided phytostabilization of mine waste. *Bio-Geotechnologies Mine Site Rehabil* 2018:147–57. https://doi.org/10.1016/B978-0-12-812986-9.00009-9.
166. Yang S, Liao B, Yang Z, Chai L, Li J. Revegetation of extremely acid mine soils based on aided phytostabilization: a case study from southern China. *Sci Total Environ* 2016; **562**:427–34. https://doi.org/10.1016/J.SCITOTENV.2016.03.208.

167. Martínez-Martínez S, Zornoza R, Gabarrón M, Gómez-Garrido M, Rosales RM, Muñoz MA, Gómez-López MD, Soriano-Disla JM, Faz A, Acosta JA. Is aided phytostabilization a suitable technique for the remediation of tailings? *Eur J Soil Sci* 2018. https://doi.org/10.1111/ejss.12727.
168. Muñoz MA, Guzman JG, Zornoza R, Moreno F, Faz A, La R. Effects of biochar and marble mud on mine waste properties to reclaim tailing ponds. *L Degrad Dev* 2016;**27**. https://doi.org/10.1002/ldr.2521.
169. Zornoza R, Á F, Martínez-Martínez S, Acosta JA, Gómez-López MD, Muñoz MÁ, Sánchez-Medrano R, Murcia FJ, Fernández Cortés FJ, López Martínez E, Espín De Gea A. Reclamation of a mine tailings pond by addition of marble waste and pig slurry for the development of aided phytostabilization. *Bol Geol Y Min* 2017;**128**. https://doi.org/10.21701/bolgeomin.128.2.010.
170. Yellishettya M, Muddb GM, Shuklab R. Prediction of soil erosion from waste dumps of opencast mines and evaluation of their impacts on the environment. *Int J Min Reclam Environ* 2013;**2**:88–102.
171. Williams DJ. Chapter 7: management of solid wastes. In: Mulligan DR, editor. *Environmental management in theAustralian minerals and energy industry*. Sydney: UNSW press in association with Australian minerals and energy environment foundation; 1996. p. 157–88.
172. Nyakatawa EZ, Reddy KC, Lemunyon JL. Predicting soil erosion in conservation tillage cotton production systems using the revised universal soil loss equation (RUSLE). *Soil Till Res* 2001;**57**:213–24.
173. Fadda S, Fiori M, Matzuzzi C. Developing rehabilitation design for the abandoned mine excavations in Central Sardinia, Italy. *Int J Min Reclam Environ* 2010;**24**:286–306.
174. Morgan RPC. *Soil erosion and conservation*. 2nd ed. London, England: Longman publisher; 1996.
175. Monterroso C, Macías F. Drainage waters affected by pyrite oxidation in a coal mine in Galicia (NW Spain): composition and mineral stability. *Sci Total Environ* 1998;**216**: 121–32.
176. Nordstrom DK, Alpers CN. Metal-sulfate salts from sulfide mineral oxidation. In: En, Alpers CN, Jambor JL, Nordstrom DK, editors. *Sulfate minerals: crystallography, geochemistry and environmental significance*. Washington: The Mineralogical Society of America; 2000. p. 305–50.
177. Jambor JL, Blowes DW. *Environmental geochemistry of sulfide minewastes. Short Course Handbook*, vol. 22. Waterloo, Ontario: Mineralogical Association of Canada; 1994. p. 438.
178. Hudson TL, Fox FD, Plumlee GS. *Metal mining and the environment*. Published by the American Geosciences Institute Environmental Awareness Series; 1999. p. 11, 41–46.
179. Pinto AP, de Varennes A, Castanheiro JEF, Balsinhas AMA. Fly ash and lime-stabilized biosolid mixtures in mine spoil reclamation. *Bio-Geotechnologies Mine Site Rehabil* 2018: 159–80. https://doi.org/10.1016/B978-0-12-812986-9.00010-5.
180. Kabas S, Faz A, Acosta JA, Zornoza R, Martínez-Martínez S, Carmona DM, Bech J. Effect of marble waste and pig slurry on the growth of native vegetation and heavy metal mobility in a mine tailing pond. *J Geochemical Explor* 2012;**123**. https://doi.org/10.1016/j.gexplo.2012.07.008.
181. Guo T, DeLaune RD, Patrick Jr WH. The influence of sediment redox chemistry on chemically active forms of arsenic, cadmium, chromium, and zinc in estuarine sediment. *Environ Int* 1997;**23**:305–16.
182. IUSS Working Group WRB. World reference base for soil resources. International soil classification system for naming soils and creating legends for soil maps. *World Soil Resour Rep* 2014;**106**. https://doi.org/10.1017/S0014479706394902.

183. Majer JD, Brennan KEC, Moir ML. Invertebrates and the restoration of a forest ecosystem: 30 Years of research following bauxite mining in Western Australia. *Restor Ecol* 2007;**15**(4):S104–15.
184. Rath BP, Das S, Mohapatra PKD, Thatoi H. Optimization of extracellular chromate reductase production by Bacillus amyloliquefaciens (CSB 9) isolated from chromite mine environment. *Biocatal Agric Biotechnol* 2014;**3**(3):35–41. https://doi.org/10.1016/j.bcab.2014.01.004.
185. Thatoi H, Das S, Mishra J, Rath BP, Das N. Bacterial chromate reductase, a potential enzyme for bioremediation of hexavalent chromium: a review. *J Environ Manag* 2014; **146**(15):383–99. https://doi.org/10.1016/j.jenvman.2014.07.014.

CHAPTER SIX

Urban Areas

Carla S.S. Ferreira[a,*], Zahra Kalantari[b], Luca Salvati[c], Loredana Canfora[d], Ilaria Zambon[e] and Rory P.D. Walsh[f]

[a]Research Centre for Natural Resources, Environment and Society (CERNAS), Polytechnic Institute of Coimbra, Agrarian School of Coimbra, Coimbra, Portugal
[b]Stockholm University, Department of Physical Geography and Bolin Centre for Climate Research Stockholm, Sweden
[c]Italian Council for Agricultural Research and Economics Arezzo, Italy
[d]Consiglio per la Ricerca in Agricoltura e l'Analisi Dell'Economia Agraria, Centro di ricerca Agricoltura e Ambiente, Roma, Italy
[e]Tuscia University, Department of Agricultural and Forestry Sciences Viterbo, Italy
[f]Department of Geography, College of Science, Swansea University, Swansea, United Kingdom
*Corresponding author: E-mail: carla.ssf@gmail.com

Contents

Advances in Chemical Pollution, Environmental Management and Protection, Volume 4
ISSN 2468-9289
https://doi.org/10.1016/bs.apmp.2019.07.004

Abstract

Urbanization and human activities create a wide variety of soils within urban environments, ranging from near natural (e.g., urban forest) to highly disturbed soils (e.g., soils in abandoned industrial areas), with great spatial and often vertical variability. Urban soils provide relevant ecosystem services, such as food production, water regulation and esthetical inspiration, but these vary depending on the intensity of landscape disturbance and degradation processes. Urban soils are highly affected by sealing, compaction (e.g. due to heavy machinery during construction activities, vehicles and human trampling) and erosion (mainly by water, particularly on bare soils, at construction sites and along road edges). Contamination from a variety of sources (e.g., inadequate waste disposal, stormwater runoff and atmospheric deposition), is often not only a major threat, but is usually very difficult to assess because of the great spatial variation of soil types, land use history, management activities and urban pattern. Heavy metals and polycyclic aromatic hydrocarbons have been widely identified in urban soils, but other chemicals also can cause contamination of urban soils (e.g., pharmaceuticals and pesticides). Biological degradation of urban soils, associated with changes in micro- and macroorganisms, affects biological activity and species richness, and can also constitute an important problem in urban areas. Depending on the degree of soil degradation, remediation (e.g., tilling and bioremediation) and restoration strategies (e.g., use of amendments, green infrastructures) might be required to enhance ecosystem functioning. Nature-based solutions are a useful approach to minimize land degradation and improve sustainable urban development.

Keywords: Urban soils; Ecosystem services; Sealing; Compaction; Erosion; Contamination; Biological degradation; Remediation and restoration; Nature-based solutions

1. Urban development and urban soils

1.1 Expansion of urban areas

Humanity is facing a broad range of challenges, including, rapid urbanization, rising population, provision of sustainable supplies of water and food, degradation and loss of natural capital and ecosystem services (ESs), climate change and associated increases in natural disaster risks.[1] By 2050, 70% of the world's population will live in cities, making urbanizing regions critical in achieving a sustainable future for the world.[2] Compactness and dispersion are the two main models under which cities have evolved.[3,4] In recent decades, the debate on compact versus sprawled trajectories of urban development has turn out to be a relevant research issue.[3] Classically, sprawl is a phenomenon associated with the rapid low-density outward expansion; it occurred first in United States cities from the early part of the 20th century, fueled by the rapid growth of private car ownership and a preference for detached houses with gardens.[5] Traditionally, urban sprawl was defined as a low density, ecologically inefficient suburban development around the periphery of cities, characterized by vehicle-dependent development occupying especially areas previously devoted to agriculture.[6] Today, urban sprawl is amongst the major concerns for its adverse environmental impact.[7] It has been argued that low-density suburban and exurban development not only degrades environmental resources (e.g., water and air quality), but also limits (or rejects) accessibility to natural resources.[7–10]

Early concerns about urban sprawl in the UK led to the creation in the 1950s of the Green Belt around Greater London and the establishment of New Towns (with green areas within them), such as Welwyn Garden City and Stevenage beyond the Belt to accommodate additional population and industrial growth of the London area. By the late 1980s, the first environmental concerns regarding urban sprawl began to appear throughout the European Union.[5] Since then, the European Commission began to work on town planning strategies that would consider, as a priority, mixed land-use and denser urban development in order to reduce the impact of sprawl on both natural and social environment.[9,10] The environmental, social and economic impacts of the phenomenon for both cities and the countryside of Europe are becoming always more evident and require immediate action, especially now that the global challenge for climate change is putting more pressure on governments.[10] For this reason, modern town planning was developed, with the objective of controlling urban expansion.[11]

In current research, the analysis of sprawl reflects urban population size, urbanization rates and land-use change. Today the diffusion of urban sprawl in Atlantic, Central, Eastern and Southern European regions has outlined mixed patterns of growth and decline combined with sprawl: Southern Europe is mainly growing with sprawl, and Northern Europe is growing with more containment.[11,12] Through data provided by the European Environment Agency,[5] some reflections on urban sprawl in Europe emerges revealing an unsustainable rate of land consumption mainly to the expense of agricultural areas.[6,8,10]

1.1.1 Urban population size, urbanization rates and land-use change

Built-up areas have grown at a higher rate than population.[13] Fig. 1 illustrates, in most European countries, changes in the percent share of built-up area over time, i.e. the ratio of built-up areas to total area of each analysis' unit. Fig. 1 uses data of the Corine Land Cover (based on 2006),

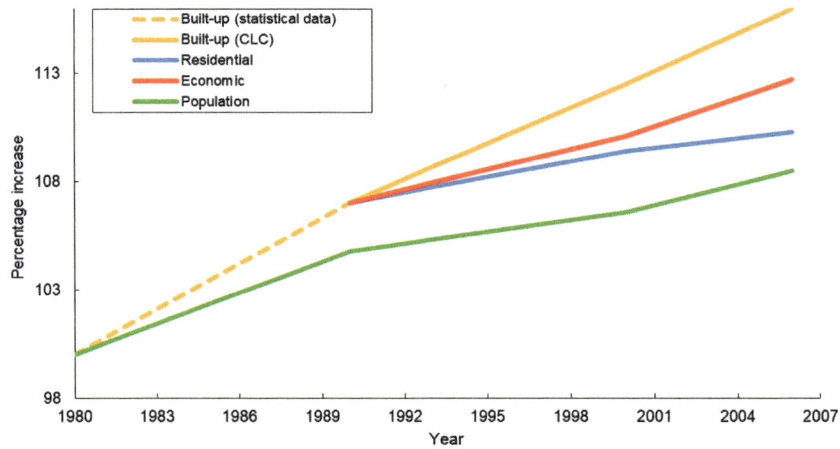

Fig. 1 Built-up area and population increase in selected countries. *Adapted from EEA.*[14]

dividing residential areas from those intended for an economic function and compares these data to demographic trends.

Land-use changes were assessed by processing Corine Land Cover maps from 1999, 2000 and 2006 (Fig. 2). Two classes of land use were considered,

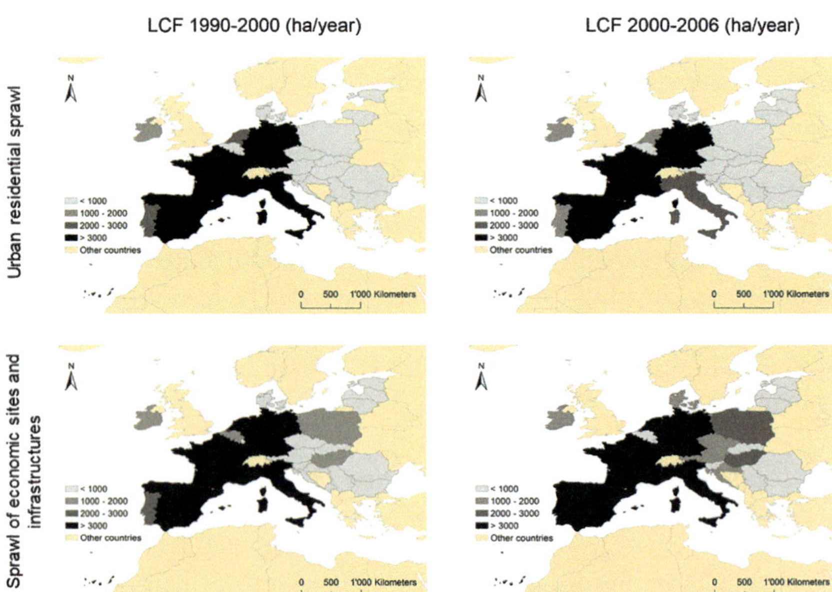

Fig. 2 Absolute change over time for urban residential and economic sites and infrastructures (ha/year) in selected countries. *Adapted from EEA.*[14]

which are those labeled as urban residential sprawl, and sprawl of economic sites and infrastructures (e.g., industrial and commercial areas). Two time periods were considered: from 1990 to 2000 and from 2000 to 2006, for selected countries, i.e. which are the majority of the European countries. Areal increase was 0.2% per year for "urban residential sprawl" and 0.3% per year for "sprawl of economic sites and infrastructures" in 1990–2000. These rates increase in 2000–6, reaching 0.9% per year for "urban residential sprawl" and 2.6% per year for "sprawl of economic sites and infrastructures".

The expansion of European cities in recent years has primarily occurred on former agricultural land (Fig. 3). Typically, urban development and agriculture are competing for the same land, as agricultural land adjacent to existing urban areas is also ideal for urban growth.[6,8] The motivations of farmers in this process are clear as they can secure financial benefits for the sale of farmland for new housing or other urban developments. Sprawling cities also threaten to consume the best cropland, displacing agricultural activity to both less productive areas and more remote upland locations with increased risk of soil erosion.[6] In addition, the quality of agricultural soils surrounding sprawling cities has also been reduced. All these characteristic impacts of sprawl are well demonstrated in the Mediterranean coastal areas.[15–17] Throughout the region, more than 3% of farmland was urbanized in the 1990s, and 60% of this land was of good (or superior) agricultural quality.

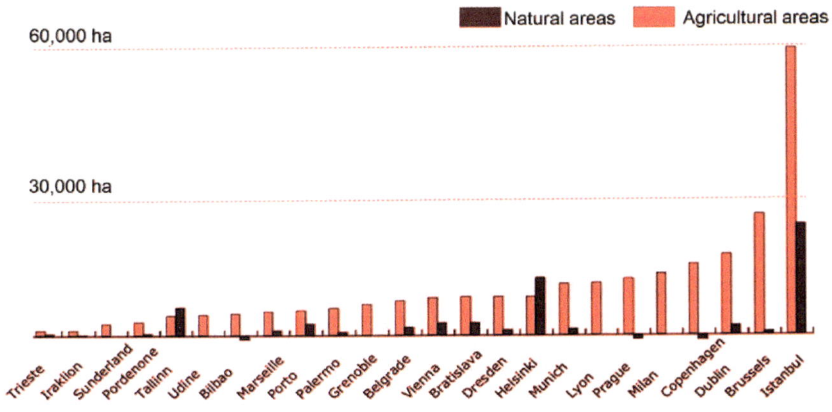

Fig. 3 Quantification of areas devoted to sprawl in pristine natural and agricultural land in selected European cities. *Adapted from EEA.*[14]

Land-use changes are also altering both water and land-surface characteristics. These conditions generally generate strong migratory flows of people looking for places offering a better quality of life.[18] Areas in southern Europe, where desertification processes are more significant, are particularly sensitive to such a situation.[19]

Highlighting the importance of transformative processes rather than the final urban outline of sprawling cities, the requirement of treating the phenomenon and differentiating it from processes of more classical (i.e. compact) urban growth has been underlined.[20] In the early 1980s, the European Commission officially expressed its concerns for the diffusion of sprawl within the continent, as it saw as contributing negatively to the accomplishment of a truly sustainable local development. Modern town planning has been unable to cope fully in reaching this objective, leading to uncontrolled expansion in many European countries over time.[10,21,22] The need for an accurate analysis of urban sprawl in Europe was becoming progressively crucial for the design of efficient territorial policies.[23] In the second half of the 20th century, Europe, and especially southern Europe, experienced a period of rapid population growth and urbanization.[13] For instance, 6 out of 10 of the European cities with the highest sprawl rates are located around the Mediterranean basin, especially in economically disadvantaged regions.[5,24]

The increasing pressure on the environment resulting from human-induced demand factors (urbanization and natural resources consumption) highlight cities which are expanding over larger areas,[4,25] thus increasing exposure to natural disasters and other events, such as land degradation, which are likely to increase both in frequency and intensity due to climate change.[9,18] This evidence supports the idea that environmental policy and regional planning should cope more effectively with the increasing vulnerability of large urban regions to natural hazards.[10]

2. Urban soils: properties and functions

2.1 Characteristics and types of urban soils

Soil is formed from disintegration of parent material by the action of climate (including water, temperature and wind) and subsequent action of micro- and macro-organisms over time, depending on topography. In urban environments, these natural soils can remain intact (e.g., patches of forest or other natural conservation area that were integrated within urban areas and remain intact), be slightly disturbed or be completely manmade. Disturbance can derive from (i) grading, excavation and compaction, as a result of construction and demolition activities (e.g., to level hillslopes, prepare soil to install building foundations or for surface sealing); and (ii) movement of soil, including exportation and addition of exogenous material, associated with relocation and infilling of low areas. Fill can be performed with (a) local soil, such as the topsoil removed prior the construction activities; (b) soils from other sites, frequently with different characteristics from natural on-site soils; (c) manufactured soils to provide best conditions for particular types of land use (e.g., gardens, parks and sport fields); and/or (d) anthropogenic materials (also called technique materials), such as construction debris (e.g., asphalt and ballast) and waste (e.g., plastic and glass). Disturbance activities help to overcome engineering and construction limitations affecting building foundations or concrete slabs for building floors, car parks, sport fields, among other land uses (Fig. 4).[26]

Soil management practices after soil disturbance, such as fertilization (e.g., compost application) and irrigation in vegetable gardens, can lead to the establishment of new and deep soil surface layers.[27] Urban soils may contain organic and inorganic contaminants derived from anthropogenic activities, from either localized or diffuse sources (e.g., industrial activity, herbicides application).[28] All these features lead to a unique combinations of factors determining soil properties and their functions.[27]

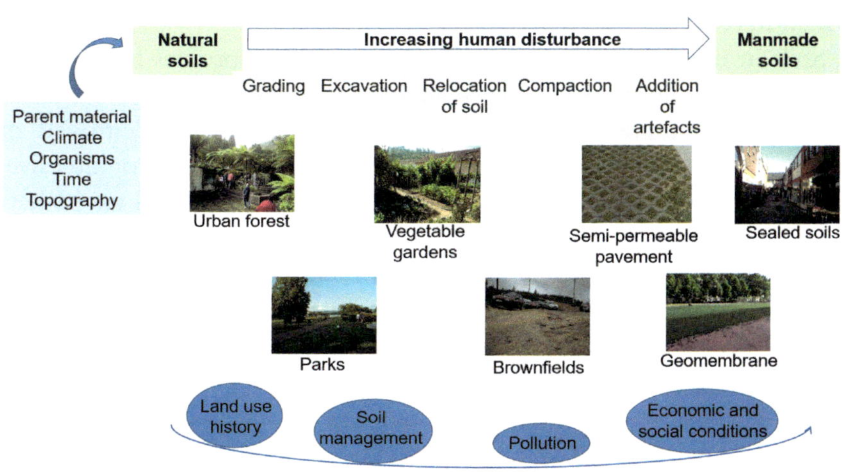

Fig. 4 Examples of types of urban soils with distinct degrees of human disturbance, driven by diverse natural and anthropogenic factors.

Urban soils (i.e. those located within urban environments) vary greatly in their properties and differ considerably from natural soils and disturbed agricultural soils of the surrounding area. Characteristics of urban soils depend on original natural soil properties, the type of disturbance activities and the nature of any added soil and/or material. Thus, a variety of uniform to very heterogeneous mixtures of particle sizes and materials are produced. This leads to a spatial mosaic of soils, driven by a diversity of soil-forming conditions, with distinct physical, chemical and biological properties, and with unique features.[29]

This large diversity makes urban soil classification and survey challenging operations[30] and special soil classification systems have been developed. Distinct taxonomic classes have been proposed, but the one from World Reference Base[30] is widely accepted. It classifies Technosols as soils dominated or strongly influenced by human-made material, comprising (i) significant amount of artifacts, which are materials made or strongly altered by humans or extracted from greater depths; (ii) sealed surfaces using technique hard material (having properties unlike natural rock); or (iii) geomembranes. Artifacts can be produced by pre-industrial processes (e.g., ceramics), containing high industrial waste (e.g., mine spoil, dredging, slag, ash, rubble), refuse of human settlements, cemented or indurated layers, or even high content of organic waste. Technosols include soils from wastes (landfills, sludge, cinders, mine spoils and ashes), pavements with their underlying unconsolidated materials and constructed soils.[31] Materials used in urban soils often contain carbonates (concrete, mortar, lime), leading to high pH values.[30]

Soils in urban environments often have different profiles from natural soils, since their horizons have been altered (by mixing the topsoil and subsoil or covering pre-existing soil layers with other soil), destroyed or removed. Disturbance in soil profile can range from low to intensive, affecting the way soil develops.[26] Great vertical and horizontal spatial heterogeneity of soil properties is one of the typical characteristics of urban soils. Abrupt changes in soil profile may be recorded, with discontinuous layers presenting significant differences in, for example, texture, structure, organic matter content, pH and bulk density, which can cause problems for plant root growth.[32] Notwithstanding these influences, urban soil geochemistry largely depends on geological, geomorphological and landscape factors, as well as the land use history, which triggers distinct cycles of change and evolution.[33] Different urban land uses, such as residential, commercial, recreational and agricultural, provide a mosaic of soil patches, which will vary in size and configuration depending on human population density, development patterns and transport networks amongst other factors.[29]

Thus, urban soils present unique challenges to landscape architects, horticulturalists, engineers, and urban planners.[26] The wide variability of urban soils, which can be noticed from a few centimeters to meters, represents a major challenge for mapping. Mapping of urban soils requires high quality and intensive soil surveys, in order to provide useful information for successful development planning.[34]

2.2 Ecosystem services provided by urban soils

Over centuries, soils in urban environments were mainly considered as simply an inevitable consequence or cost of urban development. In recent decades, however, an increasing recognition of urban soil functions and their ability to provide ESs have been noticed. ESs are the benefits that human population obtain, directly or indirectly, from the ecosystem.[35] The ability of urban soils to provide ESs are dependent on the type and intensity of anthropogenic disturbance. Unchanged or little modified urban soils can provide services similar to those in natural soils, developed under similar conditions (e.g., parent material and climate).[29] In contrast, sealed soils ensure the transportation of goods, energy, and human beings. According to the degree of disturbance, urban soils can provide services sustaining and fulfilling distinct human needs.[36]

In urban environments, the complex mosaic of different soil types provides distinct critical functions, and multiple ESs at the same time.[37] Depending on their physical, chemical and biological characteristics (natural and artificial capital), urban soils can provide several supporting, provisioning, regulating and cultural services (Fig. 5).

Supporting services are those relevant to assure the basis for the services of the three other categories. They include the support for urban developments but also habitat provision and pollination. Urban soils are mainly recognized by the support/habitat provided to soil and plant biota, sustaining biological diversity (fauna, flora and microfauna) and productivity. Green spaces, such as urban forests and gardens, have been increasingly recognized for their role in biodiversity conservation within urban areas undergoing densification.[37] Supporting services are also linked to soil formation and nutrient cycling.[32]

Provisioning services include both subsistence and commercial food and goods. Urban soils provide building materials, such as wood, organic matter and fiber, water, energy (e.g., biomass fuels) and food. Food production in

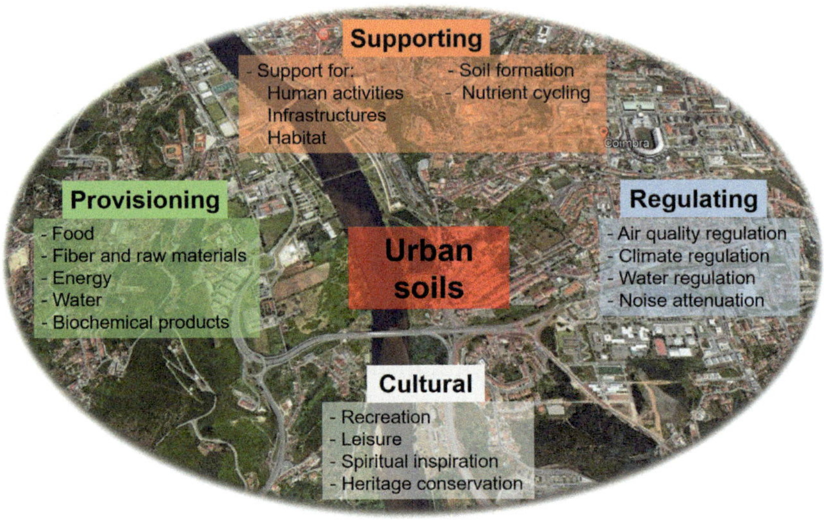

Fig. 5 ESs provided by urban soils. *Image adapted from Google Earth.*[38]

urban gardens can improve nutritional health and the local economy, and may contribute to long-term food security.[39] The relevance of urban agriculture for food security has been recognized in several countries, particularly in Asia, mainly due to economic and political changes that have undermined the food distribution systems.[40]

Regulating services include biological control on biogeochemical fluxes and climate. Pervious soils within urban environments have a high potential for carbon storage and nitrogen retention,[29] although the carbon storage capacity is strongly affected by soil type and characteristics, ranging from near zero to over 70 kg cm^{-2} at 1 m depth.[27] Carbon sequestration in urban soils helps to mitigate the effects of greenhouse gas emissions into the atmosphere,[29] and, thus, contribute to local (urban heat island) and global climate regulation.[20,36] Urban soils play an important role in water management, since they provide rainfall infiltration,[41] limiting runoff and flood hazards, and allowing groundwater recharge.[42] Urban soils also control water and air quality, by reducing the bioavailability of pollutants and providing acoustic insulation and noise reduction.[43]

Cultural services are non-material benefits that people obtain from ecosystems, such as esthetical and spiritual inspiration, recreation and cultural heritage. In urban environments, there has been an increasing demand for open spaces, such as recreation areas and children's playgrounds[27] and recognition of the value of green areas for mental and physical human health.[30]

Community gardens, typically established on derelict or abandoned areas, are also recognized by policymakers, health professionals, urban planners and the local community for improved personal wellness, environmental health benefits and sense of community.[40] Urban soils also preserve natural and cultural history of an urban area,[30] including cemetery soils (Necrosols).

The patchy mosaic of urban soils can provide multiple ESs,[44] essential for the management of environmental problems.[30] ESs provided by urban soils, however, have been poorly investigated due to the complexity of urban environments (e.g., soil cover and land use history), and the need to assess several soil physical, chemical and biological indicators. Poor information about the ESs in urban soils result in a strong lack of consideration by urban planning of the services provided.[36] Nevertheless, understanding the demands and supply of urban soil services are extremely important to attain a more sustainable urban environment. For example, a building cannot be installed in a soil with poor geotechnical properties, and a garden will not be naturally productive in a low fertile soil. Urbanization has been strongly associated with reductions of regulation services, and has a rather weak influence on cultural ESs.[44] Urban planners must counterbalance ESs supply and demand, as currently urban soils are providing very little to a whole range of services.[36]

3. Soil physical degradation

Great human pressure in soils located within urban environments often leads to their physical degradation. The European Commission highlighted sealing, compaction and erosion as key soil threats.[45] Soil physical degradation impairs the ability of the ecosystem to provide services and enhances the vulnerability of urban environments to natural hazards and climate change impacts, with impacts on communities' welfare and sustainability.[46]

3.1 Sealing

Soil sealing is characterized by the destruction or coverage of soil by layers of completely or partly impermeable artificial material (e.g., asphalt, concrete, geomembranes).[45] Sealing alters or interrupts air and water fluxes between soil and other ecological compartments (biosphere, hydrosphere and atmosphere), affecting water and biogeochemical cycles and energy fluxes.

Sealing leads to significant losses of biodiversity through reduction and fragmentation of habitats[44], inhibits or slows down nutrient cycles[45], prevents soil carbon sequestration and storage[45], and by its nature enhances traffic emissions.[45]

Sealed surfaces lead to higher temperatures than surrounding areas, since the materials used, such as concrete and asphalt, absorb more solar radiation driven by great heat capacity and thermal conductivity.[45] Research studies in Budapest (Hungary) and Zaragoza (Spain) revealed that temperatures in large sealed areas can be up to 20 °C higher than green shaded surfaces.[45] The heat island effect is one of the major welfare threats of urban communities, since high temperatures are linked with exhaustion, fainting, heat strokes, cerebrovascular disorders and respiratory problems, and affect the economy due to raising energy consumptions for cooling buildings.[47]

Soil sealing has major impacts on hydrological processes and the water cycle. Soil sealing results in lower evapotranspiration than for pervious and vegetated soils,[42] and has limited infiltration capacity when compared with similar hydrophilic soils under agriculture and forest land uses.[41] Reduced water infiltration leads to less groundwater recharge and decreased streamflow during prolonged dry periods.[27]

Reduced infiltration and evapotranspiration lead to greater surface runoff.[42] The reduced surface roughness of sealed surfaces also decreases the time of concentration of runoff.[48] This leads to flashier and larger peak discharges, increasing urban flood hazard in terms of severity and frequency.[48] Flood hazards can be further exacerbated by the limited capacity of the artificial drainage systems during heavy rainfall events, which may no longer be able to cope with increasing runoff from expanding sealed surfaces within the urban areas,[48] especially in less developed countries.

In urban environments, however, the typical patchy mosaic of different soils, ranging from unchanged or little modified to strongly disturbed or artificially created, provide a wide variety of sources and sinks of overland flow.[41] This affects the connectivity of water fluxes, the size and speed of hydrological response to rainfall and, thus, flood hazard. For example, runoff from sealed surfaces can infiltrate in surrounding pervious soil, depending on soil properties, antecedent soil moisture content and rainfall characteristics.[41] For example, surface sealing in a sandstone soil in a central Portugal study was found to produce greater increase in surface runoff than sealing in a marly limestone soil, due to inherent lower infiltration capacity of clay compared with sandy soils.[48] Furthermore, under wet soil conditions and particularly during high intensity rainfalls, limited infiltration capacity of the pervious soils may not coupe with rainfall and additional runoff from sealed surfaces.[46]

Overland flow on urban soils, however, can be also favored by limited permeability in restrictive layers over the soil profile. Restrictive layers have high density and low porosity, slowing down or even blocking vertical water fluxes, and may result from human disturbance activities, such as compaction to install building foundations and filling excavated sites with artifacts.[26] Differences in the spatial variability of urban soils affect rainfall-runoff responses, and help to explain differences in the magnitude and intensity of runoff processes in urban catchments worldwide.[46] However, the impacts of soil sealing on surface runoff and flood hazard, are influenced by catchment size.[48]

3.2 Compaction

Soil compaction is the result of external forces applied to the soil, compressing the space available between particles, and crushing the aggregates into smaller sizes, reducing pore space, particularly the macropores, and destroying the soil structure.[32] In urban environments, soil compaction is associated with (i) the use of heavy machinery during construction works[26,49]; and (ii) human and animal trampling, including footpaths[32] (Fig. 6).

Soils with high silt and very fine sand particles, coupled with low organic matter content which favor soil aggregation, are more prone to compaction.[26] Proneness to compaction also increases when soils get very wet.[45]

Soil compaction is associated with higher bulk density, reported in a Portuguese study to attain in urban areas values 30% and 60%–90% higher than agriculture and forest soils, respectively.[41] Nevertheless, a wide range of bulk density values in urban soils is normal, with relatively low values for unchanged/natural soils to high values in highly compacted soils. In Washington, D.C., bulk density ranges from 1.74 to 2.18 m/cm^3 [32], whereas in New York it ranges from less than 1.00–1.34 g/cm^3 in Central Park to 1.54–1.90 g/cm^3 in street side soils.[50] In general, residential areas tend to have less compaction than heavily urbanized sites.[51]

Compaction affects the soil ability to provide several ESs. The reduction of soil porosity strongly affects air and water fluxes. Lower porosity of compacted soils reduces both infiltration and storage capacities, leading to enhanced overland flow. In heavily compacted soils, infiltration and surface runoff can be similar to those for sealed surfaces.[52] Compacted soils also

Fig. 6 Different types of compaction in urban soils caused by: (A) heavy machinery, (B) vehicular traffic, (C) pedestrians, and (D) unplanned pedestrian paths.

enhance soil saturation by water, where the resulting waterlogged conditions produced by compaction may impair the diffusion of oxygen and aeration of the entire soil profile.[45] Soil compaction has a major impact on root development, not only because of resistance to penetration but also indirectly due to constrains in air and water fluxes (and nutrients) within the soil, determining plant growth and vigor. In compacted soils, roots may penetrate only through cracks, soil pipes or other macropores created by rotted anthropogenic materials.[32] Soil biological activity is also affected by compaction.[32]

Compacted soils tend to exhibit particular behavioral patterns, such as swelling-cracking cycles and crusting.[49] Swelling of clay soils can be a serious problem in urban areas of wet-dry climate. When soil shrinks in dry periods, cracks are formed and forces generated can cause damage to the foundations and walls of buildings, as well as to infrastructure and underground pipelines. This problem is particularly severe in areas with a seasonal climate with long and intense dry season such as semi-arid areas.[45] Crusting is more prone in bare urban soils, and can affect up to several centimeters of the surface soil.[32]

3.3 Erosion

Erosion can often be a serious problem in urbanizing areas and has become the greatest adverse worldwide environmental concern.[53] Soil erosion is a physical process characterized by the transport of soil particles recorded when external forces are higher than the cohesive forces that bind the soil together.[54] Soil erosion is triggered by wind and/or water (including rainfall and runoff), with the latter representing the most important types of erosion.[46] Soil erosion is a three-stage process, comprising detachment, transport and redistribution, and deposition of particles. Detachment occurs, for example, during rainfall events through raindrop impact (influenced by rainfall intensity), which detaches and splashes into the air or dislodges particles, and through entrainment by overland flow, if of sufficient depth and velocity, and influenced by micro-topographic patterns.[55] The transport of sediments may be of three types: (i) sheet erosion, a low magnitude process caused by uniform surface runoff over a surface that is not canalized, transporting sediments over small distances; (ii) rill erosion, triggered by concentrated runoff flowing through little streamlets, tending to become wider and deeper over the hillslopes; and (iii) gully erosion, a landform established by severe soil erosion caused by substantial runoff.[46] Landslides are severe forms of mass movement which may affect urban areas, and they are more prone in tropical regions and very steep slopes. Deposition of sediments is mostly determined by water flows over the landscape.[56]

Although erosion rates are dependent on natural factors, such as geology, geomorphology, topography (including steepness and length of slopes) and rainfall characteristics, anthropogenic factors, associated with land use and land management are of utmost importance, driven by the intensity of human disturbance.[56] In urban environments, soil erosion is most prone in bare soils, construction sites and along road edges, and due to inadequate planning of the urban drainage system (Fig. 7).[27,56] The lack of protective vegetation cover, enhanced by inappropriate management activities and advanced stages of soil degradation, makes the soil more susceptible to the effect of erosive forces.[54] Soil erosion in construction sites have been reported to attain 35–45 ton ha^{-1} per year, whereas in undisturbed woodlands it is typically less than 1 ton ha^{-1} per year.[57] Line and White[58] reported erosion rates in areas under construction about 95% greater than in forested and agricultural areas. Line et al.[59] reported sediment exports during the clearing and grading phase of a construction site nearly 10 times greater than in other land-uses, such as single-family residential areas, a golf course

Fig. 7 Examples of erosion in urban soils: (A) rill erosion in bare surfaces in construction sites, (B) gully erosion in bare surfaces under urban agriculture, (C) sheet erosion in side-street soils, and (D) pathways.

and dairy cow pasture. Ferreira et al.[60] measured rates in built-up areas in central Portugal four times higher than in disturbed forest soils. Once construction activities at the study area had finished, however, erosion rates declined to lower rates with land use stabilization.[56] However, even under stable urbanized conditions, sediment yields in urban areas are in general 6 and 3 times greater than background levels in forest and agricultural areas.[61]

High erosion rates in urban soils are produced by greatly enhanced runoff rates associated with soil sealing and compaction. These sediments can clog the urban drainage system (e.g. gutters and drains), providing overflow to surrounding areas, or be efficiently transported via urban drainage systems into river channels, lakes and reservoirs.[61] Sediment loads deposited in water bodies and hydraulic infrastructures reduce their storage capacity, enhancing the risk of flooding, and can cause constraints to navigation, and cause abrasion of hydropower turbines.[55]

In addition, deposition of sediments in water bodies can impair water quality. The presence of sediments in streamflow increases the turbidity and leads to reduced amount of light penetration, with detrimental impacts on photosynthesis, which affect dissolved oxygen concentration and food availability to aquatic life.[45] Sediments from urban soils can carry excess of nutrients (e.g., P and N), causing algae blooms and eutrophication of water bodies. Algae blooms provide constrains to swimming, fishing and navigation, but also reductions in dissolved oxygen content, threatening the aquatic species.[26] The loss of nutrients through erosion also leads to soil degradation by reducing soil fertility and impairing biomass production.[53] In the water bodies, another major concern is the impact of potential organic and inorganic contaminants adsorbed in sediment particles, such as heavy metals and polycyclic aromatic hydrocarbons (PAHs), as further developed in the next section.[62,63]

4. Soil chemical degradation

4.1 Contamination of urban soils

Soil contamination in urban environments is caused by anthropogenic disturbance activities through the addition of materials either not found in natural soils, or with concentrations above background levels.[32] Urban soils are considered to be regional sinks of pollutants,[64] due to their ability to accumulate anthropogenic chemicals continuously in the urban environments.[65] Because of this, soil contamination has been recognized as a key soil threat.[45] In 2011–12 the European Soil Data Center identified 2.5 million contaminated sites and estimated a total of 11.7 million potentially contaminated sites across Europe.[45] In Iran, increasing levels of potentially toxic elements have been recorded in urban soils.[66] In Russian Federation, 11% of the soils in medium and large cities are considered "dangerous" or "moderately dangerous" in terms of heavy metal pollution.[67] In southern China, 13% of soil samples collected in distinct urban areas displayed heavy metal concentrations considered to cause threats to environment and human health.[68]

It is difficult to assess the scale of urban soil contamination, as relatively few studies have been carried out due the to the great heterogeneity of urban soils and the difficulties in determining and mapping the exact spatial distribution of contaminants.[65] Furthermore, most studies focus only on a few pollutants, such as heavy metals and PAHs.[46] Very limited information about other urban soil contaminants is available. This is because there are more than 700 emerging pollutants, defined as synthetic or naturally occurring chemicals or microorganisms that are not commonly monitored, that have the potential to enter the environment and cause adverse ecological and/or human health effects.[45]

Urban soil contamination is prone to occur as a result of industrial, domestic and commercial activities, energy consumption and inadequate waste disposal.[46] Although largely dependent on the nature of sources, contamination is also influenced by soil physicochemical properties and landscape patterns.[66] Relevant soil characteristics include texture, organic matter, mineral composition, pH, redox potential and cation exchange capacity.[68] These soil properties will affect soil reactions with the contaminant, such as sorption, complexation and precipitation, determining its retention or mobility capacity.[29]

Urban soils typically exhibit lower organic matter content than forest and agriculture soils, due to lower biomass and litter production and sometimes removal of litter by humans. Organic matter has the ability to bind several contaminants, such as metals, so lower organic matter content in the soil increase the availability and mobility of the contaminants.[29]

Urban soils also tend to have higher pH than their natural counterparts. This can be due to (i) the application of calcium or sodium chloride for deicing, (ii) irrigation of green areas with calcium-enriched water, (iii) calcium release from building rubble comprised of masonry, cement, plaster, etc. and/or (iv) surface weathering of buildings and pavements under the acidic atmosphere and rain of urban environments. The pH has a major role in the chemical reactivity of the soil, affecting the mobility of chemical contaminants.[32]

Past history of the site is important for soil contaminant evaluation, since although time and dilution help to alleviate the danger of contaminated urban soils, some persist in toxic concentration to plants and animals for many years.[32] This is particularly relevant in evaluating the potential of unused and/or derelict urban land, for example for vegetable gardens.[40]

It is widely accepted that soil contamination within urban environments increase with the intensity of urbanization,[69] but this is not always the case. Thus, in Queensland, Australia, Goonetilleke et al.[62] reported larger pollutant loads from detached houses than multifamily dwelling units, due to a greater extent of road surface area and higher nitrogen concentrations resulting from use of fertilizers in the large gardens in the detached housing areas compared with high-density residential development. Main roads and industrial areas have been also associated with greater suspended sediments and heavy metals than residential, open space and commercial areas.[70]

Contamination of soil can affect the ability of the ecosystem to provide services such as biomass production and water supply,[68] but also can impair human health.[71] Urban populations have shown increasing interest in green areas within cities, with families, athletes and elderly people spending part of their free time in parks in particular.[72] The risk of soil contamination for humans, especially for children and seniors, is caused by direct and/or indirect uptake. Accidental soil ingestion, intake of soil particles by inhalation and dermal problems driven by skin contact, can directly affect human health.[72] This can be particularly relevant in urban soils in playgrounds, sports fields and parks through contamination by pathogenic organisms, such as dog roundworm, associated with feces of pets.[73] Soil contaminants, such as heavy metals, can also affect human heath via the food chain, through vegetables produced in contaminated urban gardens.[74] Additional studies of bioavailability of urban soil contaminants are required, however, in order to assess human health risks more exactly.[40]

4.2 Sources and mechanisms of urban soil contamination

Several human activities are able to produce contaminants susceptible to reach and accumulate in urban soils. Sources can include solid waste disposal, wastewater, industrial processes and spills, corrosion of anthropogenic material decomposition, atmospheric deposition, stormwater runoff and erosion.[27] In 2011–12, major soil contamination sources in Europe included municipal and industrial wastes (37%) and the industrial/commercial sector (33%).[45] EU Directive 2010/75 provides a list of human activities, including industrial (e.g., energy, mineral, chemical) and waste management activities (e.g., incineration), posing a risk of emitting contaminants.[75] Soil contamination by toxic elements from nearby industries such as steel manufacturing, power plants and metal smelting have been widely recorded.[72]

Dumping of waste, particularly in landfills, and the percolation of rainwater through waste layers led to soil contamination by various chemical and microbial substances. Leachate from landfill can contaminate the entire soil profile, since contaminants can move as deep as water moves into the soil, and also impair groundwater quality,[76] particularly if the underlying rocks are permeable. In highly urbanized areas, including many across Europe, groundwater is often affected by diffuse pollution from many sources.[77] Contaminants in deeper layers, however, may also be translocated upwards, as soil dries and water evaporates from soil surface. In dry areas, however, upward fluxes are restricted, as less water carry contaminants deep into the soil profile.[77]

Wastewater can be a significant source of soil contamination, particularly (a) in areas where septic tanks are used and (b) in urban areas with combined sewer systems, due to exceeding the capacity of drainage systems during larger storms.[78] Septic fields have been recognized as significant sources of nitrates, phosphate, chemical and biochemical oxygen demands, as well as coliforms.[79] Although contamination by wastewater is not a major problem in developed countries, due to the efficiency of sewerage pipe networks and treatment works, soil contamination may sometimes still occur due to leaks in the drainage system and treated discharges from wastewater treatment plants.[78]

Emissions from domestic heating, paints, industrial and power plants, as well as diverse industries and traffic are very relevant for atmospheric chemistry. The distribution of atmospheric pollutants is controlled by meteorological factors, notably wind direction and speed. Atmospheric-derived contamination of urban soils can occur via dry or wet (washed by rainfall) deposition. Topographic settings and the design of the urban areas, such as building patterns, influence wind regime and air flows.[68] For example, "canyon effects" occur along large streets surrounded by high buildings, and decreasing wind speed in closed courtyards both favor atmospheric deposition and accumulation of pollutants in the soil.[80] On the other hand, in well ventilated areas, associated with open spaces, pollutants are carried away. Thus, depending on their orientation, buildings can either protect or enhance soil contamination.[80]

Nutrients (nitrogen and phosphorous) from atmospheric deposition can be associated with tree pollen, mostly from urban forests and parks, but also lawns within residential areas,[78] and fossil fuel combustion, released from vehicle traffic and industrial activities.[76] Traffic emissions can release to the atmosphere about 10–155 mg km^{-1} of ammonia, depending on vehicle type.[81] Vehicular traffic is also associated with the emission of several heavy metals. These emissions led to 10-fold higher mass rate of metals (Zn, Cu, Cd, Ni, Pb, Hg and Ag) in residential than in open land areas of Milwaukee, USA.[82]

Runoff can be another key source of soil contamination in urban environments.[62] It is associated with a wide range of pollutants, including heavy metals,[63] nutrients,[62] chemical pollutants (e.g., pesticides and drug molecules) and fecal coliforms.[79]

Impervious surfaces have been considered as efficient transporters of pollutants, due to their efficiency in converting rainfall into runoff. Roads, and particularly highways, are major, open and dynamic sources of contaminants within urban environments.[39] Contaminants from (i) vehicle emissions, (ii) weathering and wear of road surfaces and (iii) vehicle wear (e.g., tires, brakes and fluid losses), are transferred to the soil not only through runoff, but also splash and dust.[72] The density of road networks and the frequency and intensity of traffic influence the mass transfer of contaminants to soil. Soil contamination from roads are greatest immediately adjacent to them but can extend more than 50 m.[39] Soil contamination adjacent to roads is often associated with (i) heavy metal contents (e.g., As, Cd, Cr, Cu, Pb, Pt, Pd, Ni, Sb, Zn) at levels that poses a potential risk of harm to food production, and (ii) non-threshold carcinogens (e.g., polycyclic aromatic hydrocarbons).[39] In several cities, however, decreasing Pb contamination in urban soils has been documented, as a result of the suppression of use of petrol containing tetraethyl lead.[67,83]

Rooftops can be an important source of contamination, not only through atmospheric deposition and organic residues (e.g., leaves, feces from birds), but also due to the release of chemicals from these surfaces. The type of roof material (e.g., zinc sheet, slate, interlocking tiles, flat tiles), the age and the conservation status of the roof are important parameters affecting runoff properties and pollutant loads. In Paris, France, rooftops contribute more than 80% of the Cd, Pb and Zn contamination during the wet season in the combined sewer system.[84] Runoff from sawmill rooftops along Washington coast also exceeded the water quality guidelines for Cu, Pb and Zn in all the samples tested.[85] Other studies, however, reported a low impact from rooftop runoff on urban contamination.[86]

Soils in leisure and agricultural areas within urban environments, such as gardens, lawns and golf courses, can be further contaminated with nutrients (e.g., N and P), driven by inappropriate management practices, associated with fertilizers, irrigation, weed and disease control, as well as grass clipping.[62] Herbicide and pesticide residues and their decomposition byproducts are contaminants found in urban soils, either remaining from former management practices or as residues from direct application to the vegetation.[32] In Marquette, Michigan, despite limited runoff from green areas, due to relatively high soil infiltration capacity, runoff concentrations from lawns contributed five to ten times more nutrients than other landscape surfaces, such as streets.[87]

According to Azzellino et al.[77] sources of soil contamination can be grouped into three different classes: (i) point sources, referring to localized areas releasing high concentrations of contaminants (e.g. landfills); (ii) multiple point sources, comprising areas with several point sources, each one releasing a contaminant mass (e.g. industrial areas); and (iii) non-point sources, comprising those coming from large areas. Urban runoff has been considered a major non-point source of pollutants within urban areas.[32] Nutrient losses in leachate from urban green areas, particularly home lawns, have been considered similar to forest areas, due to leachate from leaf litter, supplied by extensive tree canopy.[62]

4.3 Typical contaminants and their environmental impacts

Urban environments comprise a wide number of sources of contamination, determining the type of contaminants in urban soils. When these contaminants reach concentrations that may provide deleterious effects, such as harm to living resources or hazards to human health, they are considered pollutants. Typical contaminants in urban soils include heavy metals, PAHs, polychlorinated biphenyls, and phthalate esters/phthalic acid esters.[46] Several emerging contaminants, to date poorly investigated, are also susceptible to pollute urban soils. Emerging pollutants are categorized into more than 20 classes associated with their origins. The prominent classes in urban soils include pharmaceuticals and disinfection by-products.[45] Contaminants can be adsorbed to organic and/or minerals components of the soil and remain persistent for long periods.[65]

Urban soil contamination with heavy metals have received increasing attention because of their persistence in the environment.[67] It has been reported worldwide, as a result of vehicular traffic,[63] industrial activities (particularly energy), mining, and large non-ferrous metallurgy plants,[67] leaks from sewer systems[77] and material corrosion in urban infrastructure.[29] In recent decades, however, contamination from industry has decreased as a result of economic recession and installation of environment-friendly technologies.[67] Data from the European Environment Information and Observation Network for Soil (EIONET-SOIL) indicate that 60% of urban soil contamination is due to heavy metals and mineral oil.[45]

Heavy metals tend to be firmly fixed in soils, and so decreasing inputs do not necessarily result in declining contents in soils. There is competition between heavy metals and organic chemicals for the sorption sites in the soil, with implications for the formation of chelate complexes that have a different tendency to be adsorbed on soil compared with the free organic contaminants.[28] Thus, urban soils can act as both a sink and a source of heavy metals.[40]

Soil contamination with heavy metals can have adverse ecological effects, through their impact on soil microorganisms and disruptions of biogeochemical cycles, by impacting both the physiology and ecology of organic degrading microorganisms,[28] changes on foodweb functioning, and the potential risks to plants and human health.[46,67] Although some heavy metals, such as Cu and Zn, are essential for living organisms, studies have demonstrated adverse influences of high bioavailability of heavy metals for plants, including (i) damages to roots and their development, and (ii) leaf problems, such as chlorosis, brownred coloration and deformation of young leaves.[72] Heavy metals may also lead to human injuries, particularly on nervous, circulatory and immune systems,[68] with strong negative impacts on the quality of life and decreasing life expectancy.[33] Lead has been considered the most deleterious heavy metal, due to high toxicity and neurotoxicity, and possible human carcinogenic properties. High doses of Zn may cause dermatitis, skin irritation, neuropsychic dysfunctions and necrotic effects on nervous cells. Cd is a highly toxic metal, with carcinogenic, mutagenic, genotoxic and immunotoxic effects. Exposure to Cu may cause liver and kidney damages, digestive problems, mucosal and central nervous system irritation.[72]

In soils from urban lawns, heavy metals can interact with each other and with agrichemical contaminants, enhancing their individual toxic effects.[28] Pesticides are chemicals used to reduce pest damage by suppressing or repelling them. They include herbicides used to eradicate plants, insecticides to destroy insects, and fungicides to surpass fungal organisms.[57] Herbicides are widely used during planting and maintenance of leisure green areas. Siderum, for example, a widely used pre-emergence herbicide, accumulates primarily in the topsoil layer (0—20 cm), reaching concentrations ranging from 1.13 to 107 mg kg^{-1} in residential turf and golf course greens.[28]

Some urban soils are known to be contaminated with PAHs, at levels that sometimes cause potential hazards.[64] PAHs are ubiquitous organic contaminants, containing only carbon and hydrogen of multiple membered rings. Due to their volatility, PAHs may be transported far from their original sources and deposit in soil surface. Generally, dating of these compounds confirms the accepted assumption that accumulation of PAHs in urban areas is mostly related to a long-term deposition of contaminated particles.[64]

PAHs in urban soils are mainly derived from thermal alteration or incomplete combustion of organic carbon.[88] PAHs are classified as either (i) pyrogenic, when derived from incomplete combustion of organic carbon under high temperatures, such as in vehicle emissions; (ii) petrogenic, originating from petroleum products such as coal, crude oil and gasoline, at

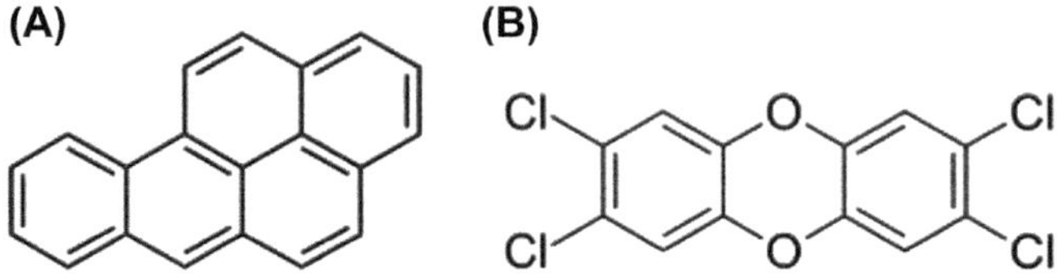

Fig. 8 Chemical structure of (A) benzo[a]pyrene, the most toxic compound between the PAHs; and (B) 2,3,7,8-tetrachlorodibenzo[p]dioxin, the most toxic compound from PCBs.

relatively low temperatures; or (iii) biogenic, produced during degradation and transformation of organic carbon by plants and microorganisms. Some studies report that soils within populous cities are mainly contaminated with PAHs of high molecular weight of pyrogenic origin.[85] Among several PAHs, the 4-ring compounds tend to accumulate more in earthworms, since they do not volatilize or degrade in the soil.[64]

PAHs are persistent organic contaminants, and some of them, such as benzo[a]pyrene (Fig. 8A), are toxic, carcinogenic (links with skin, lung, bladder, liver and stomach cancer have been reported), and affect the cardiovascular system.[89]

Polychlorinated biphenyls, a group of synthesized organic compounds with chlorine atoms and benzene rings, are classified as persistent organic pollutants. Because of their strong adsorption to soils and persistence, as well as their teratogenic (associated with lower cognitive capacity and motor control problems), immunogenic (e.g., endocrine disruptor, neurotoxic) and probable carcinogenic health effects,[90] Polychlorinated biphenyls production has been banned in several countries since the late 1970s. Polychlorinated biphenyls were widely used as dielectrics in transformers and large capacitors and can still be found in old polychlorinated biphenyls-containing, electrical equipment, leading to soil contamination in urban areas.[91] One of the most toxic polychlorinated biphenyls is the 2,3,7,8-tetrachlorodibenzo[p]dioxin (Fig. 8B), produced from waste incineration, metal production, and fossil-fuel and wood combustion.[92]

After decades of excessive use of antibiotics, their prevalence and persistence in soils has become an emerging environmental issue, particularly in areas undergoing rapid urbanization.[83] Urban soils can be contaminated via industrial and municipal waste discharge, wastewater and urban runoff. Nearly 90% of the antibiotics consumed by humans are excreted in urine and feces, as parent compounds or as byproducts, and out-of-date prescriptions are typically discarded directly into the sewage systems or in garbage cans.[83] Several types of antibiotics have been detected in wastewater treatment plants worldwide, whose treatment is not sufficient to completely eliminate pharmaceutical compounds. The biocide activity of antibiotics can have deleterious effects not only on soil microorganisms, decreasing bacterial diversity and changing dominant species, but also threaten global public health.[83]

Soils near industrial activities are often contaminated with organic compounds, such as chlorinated solvents, tetrachloroethylene, trichloroethylene and trichloromethane, and/or inorganic compounds, e.g. hexavalent chromium, substances.[77] All these compounds comprise features that favor their persistence in the environment and toxicity. Trichloromethane is suspected of being a human carcinogen under unfavorable environmental conditions, while tetrachloroethylene and trichloroethylene may undergo reductive microbial dechlorination to dichloroethene (DCE) isomers and vinylchloride, which are highly carcinogenic compounds. Soil contamination by tetrachloroethylene/trichloroethylene and trichloromethane have been reported in continental Europe, North America and other industrialized areas of the world.[77]

5. Soil biological degradation

In the soil there is a complex web of microorganisms, and it is not an easy task to understand the mechanisms, processes and attributes that make unique microbes. Given the complex and large microbiomes, the soil is a well-established reserve of biodiversity, providing a wealth of new insight into huge diversity of microbial species. The large genetically and physiologically variability of bacteria, fungi, and archaea is due to climatic, physical, ecological, and structural soil differences, and to soil management practices. In addition to the biological differences that make understanding of the microbial ecology rather challenging, some artifacts introduced into the soil make the study of urban microbes even more complex. Microbial communities' assemblages and patterns of co-occurrence and co-exclusion vary at spatial and time-related scales. The perception of environmental variability and the scale at which specific dynamics and characteristics, are measured can misrepresent the spatial scale at which microbial groups differentiate their structure.[93,94]

Microbial diversity represents the major source of biodiversity with an estimate of several million species, the vast majority of which (perhaps 98%–99%) are still to be isolated and classified.[95] It has long been recognized that soil biological diversity is the foundation of the complex ecological processes and for the maintenance of ESs. Soils provide a large number of ESs that are of pivotal importance to "sustainable agriculture", to all known food productive chains, to human life (including air quality regulation, climate regulation, water regulation, water purification, disease regulation, natural pest control, pollination, erosion control, cultural services). The primary services include nutrient cycling, regulation of water flow and storage, regulation of soil structure and maintenance, regulation of other biota, regulation of abiotic stress, and climate change mitigation.[96]

However, threats such as population and human growth, climate change, urbanization, agricultural practices, erosion and compaction, nutrient depletion, loss of soil organic carbon and ever-increasing demand for food are leading to reduced functioning of soil ESs and are resulting in the decline and simplification of biological diversity. Here the term "decline" refers to a degradation of soil quality at local level. Usually, soil biological decline can be linked also to erosion, organic matter depletion, salinization, compaction and contamination. Several factors affect the inventory diversity within the soil, but microbes are sensitive to disturbances from several threats. The soil microbiota has its own unique capacity to resist events that cause disturbance or change and a certain capacity to recover from these perturbations. The capacity to recover from change and to adapt to the present state by altruistic, cooperative and co-occurring behavior is considered a key attribute of microbial community intelligence.[97] This suggests that as soils and soil biodiversity are globally important, hence it is time to know not only this hidden belowground diversity, its composition and shape, but also how changes in soil biological diversity may affect ecosystem functioning. No environmental policy can be defined and enforced without a precise knowledge of the actual situation, of its evolution and without monitoring its effects in all habitats including those where humans spend their lives, sharing the 'good, the bad and the ugly' microorganisms. The many studies linking microbial diversity to human health, are focusing on anthropogenic influences on microbial diversity, including urban microbiomes.[98] Microbial diversity in agriculture is a special example of interaction between environmental conditions, microbes and man's activities. Microbial communities involved in food farming are strictly related and are influenced by many genomic interactions and food competitions throughout the entire farm and food supply chain. But this is true also for urban microbiomes, where stresses are predicted to reduce diversity with consequences directly linked to human health and resistance.[98] The fact that urban areas are rapidly expanding with different degrees of environmental stresses, poses a major threat to biodiversity and its decline as well as to simplification.[99] In the scenario of global change, climate change together with land use and increasing urbanization have been shown to affect soil biological diversity. Hence, decline of soil biological diversity is a matter of concern not only in agroecosystems and in natural ecosystems, but also in urban and man-habited areas.[98]

Most studies to date have focused on soil microbial communities of agroecosystems and natural ecosystems and have shown implications for biodiversity and their function, due to agricultural intensification and land use. However, to date, our knowledge of the urban microbiome, and on the effect of urbanization on soil biological diversity, is poor. If the environment and human microbiome are interrelated[100] the decline of soil biodiversity, including microbial communities, raises the question of its implications for ecosystem functioning and human health.[101] In a recent study on microbiome of built environment and surfaces of environment within clinical building, Mahnert et al.[98] have documented that the loss of microbial diversity correlates with an increase in resistance, which call for a strategy to restore bacterial diversity in certain environments where people spend their life. Following the "biodiversity hypothesis" (according to which a reduced contact of people with natural environment features and biodiversity may adversely affect the human commensal microbiota and its immunomodulatory capacity), Reese et al.[99] unraveled the relative microbial diversity of soils of urban habitats affected by different degrees of stress (small, high, low, higher), using a direct comparison with macroorganisms (insects and ants). They highlighted that diversity and composition of microbial communities are less sensitive to patch habitats investigated, and to urban stress than macroorganisms, highlighting the higher resilience of microbes and their associated ESs and functions, well recognized as "microbial intelligence". This indicate that the scale is important in community ecology of microorganisms. Ramirez et al.[102] showed that the bacterial diversity patterns of Central Park, within a highly urbanized area, are not different from that observed across natural ecosystems. Xu et al.[103] discussed the impact and implications of urbanization on bacterial community composition in urban park soils, showing that microbial composition is mainly driven by geographical locations and soil properties. Yan et al.[104] investigated the relationship between the composition and diversity of microbial communities in urban soils, aiming at understanding of bacterial diversity and abundance, to determine which factors have an important effect on the composition and shaping of microbial communities. They concluded that urban development can alter the composition of the soil bacterial communities. The mechanisms underlying these implications remain to be clarified, but they are profound and given the high dispersal potential of many microorganisms, the interactions between soil microbial communities and urban areas, are likely to happen over multiple spatial and temporal scales. Increased urbanization and urban development may have implications in terms of loss of soil biological diversity, depending on features and stress of different urban areas, which have also implications for human health.

6. Remediation and restoration strategies

Degradation in urban areas is a major problem of growing social and political concern, particularly due to the multiple uses supported and the impacts for human wellbeing. Understanding the problems inherent to degraded urban soils is essential to enhance ESs. The complexity of urban environments (e.g., population density, infrastructures, soil contamination), however, brings additional challenges for remediation (e.g., removal of pollutants) and restoration (enhancement of ecosystem functions) strategies, used not only to improve the ecological condition, but also to provide a suite of co-benefits to communities.[105]

Improving soils is key to enhance the overall ecosystem functioning in urban areas.[40] City planners and community organizations have been demonstrating increasing interest in unused urban land worldwide (e.g., resulting from the deindustrialization of metropolitan areas), since it represents available land to, e.g., produce food, manage runoff and mitigate the urban heat island effect.[40] Nevertheless, the use of this land may require the application of remediation and restoration techniques to guarantee public health. In Europe, the cost of management of contaminated sites is estimated around 6 billion Euros per year.[45] In the United States, stream restoration exceeds a billion US dollars annually.[61]

6.1 Remediation techniques

Compaction in urban areas can be tackled via tillage strategies. Tillage operations, involving turning over or mixing soil layers (e.g., ripping or scarification), may be required to enhance soil structure and mitigate compaction. This can be relevant for vegetable gardening areas, but increasing susceptibility to soil erosion also needs to be considered.[51]

With contaminated and polluted urban soils, self-cleaning can take many years. Therefore, depending on the degree of contamination, soil replacement or remediation may be required.[67] Excavation and removal activities of contaminated soils or structures are extremely expensive and even prohibitive if required on a large scale. Thus, in some cases, soils are just covered with a clean soil cover, without prior soil removal. Nevertheless, obtaining required amounts of clean soil to cover large areas of urban degraded soils is challenging. The reuse of clean sediments extracted from urban construction sites to lower the toxicity of urban topsoil can be a sustainable solution.[71]

Bioremediation involves long-term strategies for remediation of contaminated soils, using plant, microorganisms and/or soil amendments to improve the chemical properties of the urban soil.[51] Phytoremediation is a type of bioremediation usually applied in less contaminated sites. It is classified according to the contaminant removal mechanism into: (i) phytostabilization or phytoimmobilization, based on plant ability to decrease the mobility and/or bioavailability of a contaminant, through adsorption by roots, precipitation or complexation in the root zone; (ii) phytovolatilization, where contaminants are converted into volatile forms and released into the atmosphere through stomata; and (iii) phytoextraction, through chemical storage in their roots, stems or leaves, often involving harvest of the plant material to remove contaminants from a site. In phytoremediation, the microorganisms associated with the rhizosphere of the plants play a fundamental role in minimizing the toxic effects of contaminants. It is an economically feasible and efficient technique used, for example, for remediation of soils contaminated with heavy metals.[106]

6.2 Restoration practices

Depending on the aim and ultimate use of the urban soil, several techniques can be implemented to restore soil ESs, including (i) use of amendments to improve degraded soils that are unable to support any functional ecosystem; (ii) recovering nutrients, especially phosphorous, which may be used in gardening activities; and (iii) green infrastructures to enhance, e.g., water infiltration.[40] Soil amendments are materials introduced in the soil to modify specific soil characteristics, and can be classified into five types, as presented in Table 1.

In urban areas there are a wide variety of municipal solid waste (e.g., plant residues and kitchen waste) that can be composted and used to improve the fertility and ecosystem function of urban soils. Additionally, a significant proportion of nutrients can be harvested from domestic wastewaters, given their high levels in human feces and urine. Depending on

Table 1 Types of amendments used in urban soils and their benefits.

Type of soil amendment	Materials involved	Benefits for soil
Organic	Organic material (e.g. composted residues and sludge) that increase the organic matter content of the soil	- Improve nutrients content and food web - Enhance chemical buffering - Improve physical properties, such as decrease bulk density and increase water infiltration and aeration - Stimulates soil biological activity - Increases total carbon stock
Mineral	Natural (e.g. gravel and sand) or processed mineral products (e.g., perlite, hadite and pumice), chemically stable and sterile, used to change the texture of the soil	- Increase soil drainage capacity - Stimulates secondary biological activity - Absorption of a wide array of chemicals
Physical	Manufactured additives (e.g., grids, geowebs, and turf cells) designed to amend natural soil structure (commonly used to stabilize soils for grasses and small plants)	- Improve soil resistance to compaction - Adjust drainage characteristics - Ability to control erosion
Biological	Condensed organic compounds (e.g., kelp extracts) and biological inoculants (e.g., mycorrhizae, micro- and macro-fauna) to change soil biological and chemical characteristics	- Recover soil structure and porosity - Improve the soil food web - Change the bioavailability of contaminants
Chemical	Compounds used to affect soil chemistry (e.g., fertilizer), usually associated with nutrients and pH adjustments	- Correction of chemical imbalance

Adapted from EPA, 2011.

wastewater treatment level in each plant, effluent discharged can be used for irrigation. Recovery technologies can be also used to appropriate nutrients, via struvite and magnesium-ammonium-phosphate fertilizers.[40]

Green infrastructures are natural or engineered systems that mimic natural processes, particularly to manage stormwater runoff. These sustainable drainage systems, using soils and vegetation, are able to collect, provide retention or filtration of runoff pollutants, and infiltrate runoff from surrounding paved surfaces. These bioretention systems might need modified soil layers to promote increased infiltration and to support the desired plants.[51] Examples of these systems include infiltration trenches, detention basins, rain gardens and vegetated swales. Green infrastructures contribute to several ESs, and reduce the frequency of overflows of typical urban drainage systems.

7. The role of nature-based solutions for sustainable urban development

In an increasingly urban world with long-term land-use changes (notably, forested areas and grasslands being converted to residential, commercial or industrial areas), human systems (economic and land use planning) and natural systems (land use limitations) need to be considered when striving for sustainable urban development.[107] Urbanization and climate change put people in direct conflict with nature. As urbanization proceeds and population centers expand and become more interlaced with agrarian landscapes, ecosystems and the services they provide are pushed to the limits.[108] These changes in environmental conditions can have enormous consequences for the people and the places affected.

Climate change effects will accelerate and exacerbate the societal impacts of environmental changes, now and into the future.[109] Therefore, changes in climate and hydrology (and associated biogeochemistry and ecology) not only have direct relevance for societal health and the economy, but also have indirect implications for natural processes, ecosystems and their services.[110] Without a clear understanding of the feedbacks between climate change and population expansion and the resulting changes in ES, it is impossible to work successfully toward sustainability.[111] Climate change is expected to substantially alter temperature and precipitation regimes, leading to more intense and more frequent extreme rain events, droughts, reduced air quality and consequently negative influences on biodiversity and ecosystems in cities.

The vulnerability of populated areas to natural disasters is partly a consequence of decades of spatial planning policies that have failed to take full account of hazards and risks in land use zoning and development decisions. Another factor is unsustainable management of natural resources, the result of policies that focus on only one ecosystem service, without understanding the wider impacts on the natural capital underpinning ecosystem functioning. Water-based natural disasters are a typical example of the conflict between urbanization and the environment. Floods and droughts in urban areas have enormous environmental, social and economic impacts, with single events often causing damage costing millions of euros.[112]

Adapting to urbanization and climate change thus requires a change of paradigm from relying exclusively on defensive action against natural hazards to combining defensive action with adaptive management of natural resources.[113,114]

As a result of rapid urbanization worldwide, urban ecosystems are becoming the major ESs provider for humanity. Cities are a frontier for urban ecological research, requiring a synthesis of socio-cultural and ecological approaches. They can be conceptualized as complex adaptive systems, consisting of novel ecosystems with much internal variability due to a multiplicity of dynamically interacting factors. This requires the use of multiple methods and tools, and of cross-cultural comparisons, to identify general patterns that may be significant to the ecology of cities, from more specific dynamics driven by local factors of history, politics and culture.[108]

After decades of neglect, the importance of protecting and restoring urban-ecosystems has started to receive attention in recent years.[115] However, the major challenge with complex topics such as restoring urban ecosystems and climate change adaptation is that there is no single solution that fits every environmental, regulatory, economic and social context.

In efforts to cope with the threat of land use change due to population growth and climate change, the European Union (EU) has introduced nature-based solutions (NBSs).[1] Such solutions represent a better way, beyond conventional concepts, that consider both natural function and human application, and thus fit with the emerging concept of Ecological Civilization Construction. In order to scale up restoration of ecosystems as a way, for example, to fight climate change, the UN declared 2021–30 a "Decade of Ecosystem Restoration",[116] focusing on four main goals: Sustainable cities and communities, Climate action, Partnerships for the goals, and Good health and wellbeing. Ecosystem Restoration is the "process of assisting the recovery of an ecosystem that has been degraded, damaged or destroyed".[117] Having an EU agenda on NBS will enable Europe to become a world leader in the growing market for NBS.[1]

In general, NBS involve conservation, enhancement and use of biodiversity by society in a sustainable manner, while also integrating social factors such as socio-economic development and effective governance.[114] The European Commission defines NBS as solutions that are "inspired and supported by nature, which are cost-effective, simultaneously provide environmental, social and economic benefits and help build resilience". NBS thus have clear links to concepts such as ecosystem-based adaptation to climate change and green infrastructure (GI), a tool for providing ecological, economic and social benefits through natural solutions. All these approaches can help reveal the value of the benefits that nature provides to human society and mobilize investments to sustain and enhance them.[118] Nature based solutions combines existing approaches of ESs, green-blue infrastructure, ecological engineering, ecosystem-based management and natural capita with systemic solutions that combine technical, business, finance, governance, regulatory and social innovation.[1]

In practice, NBS focus on working *with* (rather than against) nature and, in essence, aim at increasing the natural capital of ecosystems, for example reducing urban flood risk or increasing resilience to drought. The underlying assumption in NBS is that hazards can be measured in terms of environmental, social and economic impact.[119] To be successful, NBS must take into account local social-ecological systems, so that city communities are engaged, in order to secure the thresholds of interventions that can effectively reduce the vulnerability of cities.[120] All the approaches on which NBS builds (ecosystems, ESs, ecosystem-based adaptation/mitigation, green-blue infrastructure) recognize the fundamental importance of working with nature and of using a systemic and holistic approach to environmental change based on an in-depth understanding of the structure and functioning of ecosystems and the social and institutional context within which they are situated.

Land degradation in urban areas is one of the current societal challenges for which NBS can play a role in recovery, due to the closed link between the natural processes in ecosystems.[121] The NBS applied in land restoration and management tasks can be flexible and provide multi-benefit types, e.g. preventing, mitigating and adapting to climate-related risks, enhancing carbon storage and more generally improving human wellbeing and resilience to global change.[1] This paves the way toward a more resource-efficient, competitive and greener economy.[107]

Nature based solutions are already beginning to reframe discussions and policy responses as countries implement measures to limit extreme events (and their impacts) and adapt to climate change. Nature based solutions for urban areas are in general large-scale initiatives that promote coordinated development and management of actions regarding water, soils, restoration of ecosystems and related resources. Nature based solutions in urban areas can be seen as ways to make anthropogenic landscapes more nature-compatible, with more new biodiversity hotspots and less land degradation.[122] The scale of NBS and their need for constant maintenance create long-term work demands, mostly at city and community level. However, it is important to stress that NBS do not compete with conventional engineering-based approaches, but rather complement and support them — or in some cases reduce the need for them. Thus, implementation of NBS creates new market opportunities without excluding existing opportunities. Nature based solutions in urban environments provide much greater benefits to citizens than previously acknowledged.[56] Building more robust evidence about the benefits of NBS will help justify investment in their implementation.

One common form of urban NBS is to have street trees, parks and urban green areas providing a range of natural benefits. These include intercepting dust, toxins and noise, sheltering and cooling buildings, acting as a carbon sink, buffering flooding, etc. More complex urban NBS provide further benefits, such as increased biodiversity, species conservation, energy production and waste management, while promoting social cohesion using collaborative processes. This requires comprehensive co-design and co-creation of ideas, leading to multiple ecological, environmental and social gains.[105]

Wetlands can be seen as possible large-scale NBS, supplying and restoring ES to mitigate societal and environmental challenges such as water pollution, flooding and drought.[123] Wetlands also have the capacity for dampening and regulating water flows, improving water quality, increasing the carbon sequestration and landscape attractiveness.[123]

There are various NBS that can be used to reduce soil erosion, such as straw mulches[124] and vegetative filter strips.[125] Natural vegetation cover can be also used as a NBS to restore degraded man-made ecosystems. For instance, green roofs can be an important NBS in urban landscapes to reduce urban stormwater runoff by increasing shoot biomass and evapotranspiration rates.[126]

The sustainable management, conservation and restoration strategies are vital to make communities more resilient to climate change impacts.[127] For example, coastal restoration and floodplain restoration help at moderating sea level rise, flood and droughts.[1] More specific, floodplain restoration can be multi-beneficial by providing flood protection, improving living condition for species, restoring carbon-rich soils, reducing nutrient loads increasing recreational facilities and boosting the economy.[128] Particular attention should be paid to cooperation between society, local communities, urban planners, architects, construction companies, etc., in restoration actions. This involves re-connecting people with nature, increasing societal awareness to implement NBS in urban ecosystems.

To achieve long-term sustainable solutions to environmental and socio-economic challenges, greater efforts are needed in ecosystem restoration and to develop and advance effective NBS.[120] Implementation of efficient NBS in urban landscapes requires a more comprehensive understanding of coupled natural and human systems, including causal mechanisms, impacts, and feedbacks.[112] It is then essential to understand barriers to and opportunities for efficient NBS, the social embedding of best management practices for these, and policy and regulatory frameworks that can drive their implementation.[129] With a clear understanding of NBS interventions to specific environmental and societal challenges, it will become possible to scale up implementation of NBS and strengthen their impact in meeting environmental and socio-economic challenges in land management and development.[107]

By mainstreaming NBS in urban planning, landscaping and regional policies, one can work toward providing solutions to safeguard urban areas, adapt to climate change and improve ESs. There is international acknowledgment that efforts to reduce disaster risks must be systematically integrated into policies, plans and programmes for sustainable development, economic development and poverty reduction, and supported through bilateral, regional and international cooperation, including partnerships.[130] However, adequate proof-of-concept of the environmental, social and/or economic benefits provided by NBS is still needed, to promote their inclusion in planning and decision making.

8. Final considerations

The expansion of the built environment brings a huge pressure on soil resources. Urban soils are subject to major changes, such as incorporation of artifacts, which affect their natural properties. The characteristics of urban soils vary according with the nature and history of disturbance, the management practices and the effect of urban environmental changes. The multiple land-uses within urban environments and different degrees of disturbance within them lead to great vertical and spatial changes and differences in soil properties, making it impossible to define or describe a typical urban soil. Although urban soils affected by intense human pressure present distinct physical, chemical and biological properties compared with surrounding natural soils, they still exhibit a surprising capacity to provide ESs to inhabitants of urban areas and settlements, depending on their degradation status.

Characterization of the urban soils and evaluation of land use suitability should be a first step for urban planners and decision makers to consider in sustainable land use planning. Although mapping urban soils is a major challenge, due to significant local variations in soil properties over small distances, detailed soil surveys are important to evaluate their ability to provide required functions. This information would help local authorities and stakeholders determine the best land-uses and management practices, with positive impacts for human health, quality of life and wellbeing. For example, urban vegetable gardening should not take place on contaminated soils so as to avoid a soil-to-plant-to-human exposure pathway.

Remediation and restoration activities, however, might be necessary to improve soil properties in order to supply the desired ESs, despite the economic costs involved. Depending on soil properties and the degradation status of the soil, tillage operations for de-compaction, bioremediation practices to decrease soil contamination, application of soil amendments, such as compost, to improve soil structure and fertility, and installation of green infrastructures to enhance water infiltration and retention, are some of the strategies that can be used in urban soils. Nature-based solutions can bring additional benefits to the urban areas, improving the overall environment and living conditions of urban communities. For example, maintaining good soil structure and vegetation cover can enhance soil structure and stability, increase drainage and water-holding capacity and reduce pollution of surface waters. They also provide additional services such as climate change mitigation, sustain biodiversity and recreation.

Using information about soil quality to support urban planning strategies would contribute to mitigate land degradation and to develop sustainable and resilient urban areas, by optimizing ESs.

Acknowledgments

This research was supported by the Portuguese Science and Technology Foundation, through the Post-Doctoral Fellowship SFRH/BPD/120093/2016.

References

1. EC. *Towards an EU research and innovation policy agenda for nature-based solutions & Re-naturing cities*. Directorate-General for Research and Innovation. 2015. Climate Action, Environment, Resource Efficiency and Raw Materials EN (full version). 2015.
2. United Nations, Department of Economic and Social Affairs, Population Division. *World urbanization Prospects: the 2014 Revision, highlights* (ST/ESA/SER.A/352). 2014.
3. Bruegmann R. *Sprawl: a compact history*. Chicago: University of Chicago Press; 2005.
4. Schneider A, Woodcock CE. Compact, dispersed, fragmented, extensive? A comparison of urban growth in 25 global cities using remotely sensed data, pattern metrics and census information. *Urban Stud* 2008;**45**:659–92.
5. EEA: European Environmental Agency. *Urban sprawl in Europe*. Copenhagen: EEA Report; 2016.
6. Cecchini M, Zambon I, Pontrandolfi A, Turco R, Colantoni A, Mavrakis A, Salvati L. Urban sprawl and the 'olive' landscape: sustainable land management for 'crisis' cities. *Geojournal* 2019;**84**(1):237–55.
7. Frenkel A, Ashkenazi M. Measuring urban sprawl: how can We deal with it? *Environ Plan B* 2008;**35**:56–79.
8. Colantoni A, Zambon I, Gras M, Mosconi E, Stefanoni A, Salvati L. Clustering or Scattering? The spatial distribution of cropland in a metropolitan region, 1960–2010. *Sustainability* 2018;**10**(7):2584.
9. Marchetti M, Vizzarri M, Lasserre B, Sallustio L, Tavone A. Natural capital and bio-economy: challenges and opportunities for forestry. *Ann Silvicultural Res* 2014;**38**(2): 62–73.
10. Salvati L, Zambon I, Chelli FM, Serra P. Do spatial patterns of urbanization and land consumption reflect different socioeconomic contexts in Europe? *Sc Total Environ* 2018;**625**:722–30.
11. Biasi R, Colantoni A, Ferrara C, Ranalli F, Salvati L. In-between sprawl and fires: long-term forest expansion and settlement dynamics at the wildland–urban interface in Rome, Italy. *Int J Sust Dev World Ecol* 2015;**22**(6):467–75.
12. Hasse JE, Lathrop RG. Land resource impact indicators of urban sprawl. *Appl Geogr* 2003;**23**:159–75.
13. Zambon I, Rontos K, Serra P, Colantoni A, Salvati L. Population dynamics in southern Europe: a local-scale Analysis, 1961–2011. *Sustainability* 2019;**11**(1):109.
14. EEA: European Environmental Agency. *Mapping guide for a European urban Atlas*. Copenhagen: EEA Report; 2010.
15. Cuadrado-Ciuraneta S, Durà-Guimerà A, Salvati L. Not only tourism: Unravelling suburbanization, second-home expansion and "rural" sprawl in Catalonia, Spain. *Urban Geogr* 2017;**38**(1):66–89.
16. Duvernoy I, Zambon I, Sateriano A, Salvati L. Pictures from the other side of the fringe: urban growth and peri-urban agriculture in a post-industrial city (Toulouse, France). *J Rural Studies* 2018;**57**:25–35.
17. Di Feliciantonio C, Salvati L. 'Southern'Alternatives of urban diffusion: Investigating settlement characteristics and socio-economic patterns in three Mediterranean regions. *Tijdschr Econ Soc Geogr* 2015;**106**(4):453–70.
18. Craglia M, Leontidou L, Nuvolati G, Schweikart J. Towards the development of quality indicators in the 'Digital' city. *Environ Plan B* 2004;**31**:51–64.
19. Salvati L, Zitti M. Land degradation in the Mediterranean basin: linking bio-physical and economic factors into an ecological perspective. *Biota* 2005;**5**:67–77.
20. Couch C, Leontidou L, Petschel-Held G. *Urban sprawl in Europe-landscapes, land-use change policy*. Oxford, UK: Blackwell Publishing; 2007.

21. Corona P. Forestry research to support the transition towards a bio-based economy. *Ann Silvicultural Res* 2015;**38**(2):37–8.
22. Salvati L, Serra P. Estimating rapidity of change in complex urban systems: a multidimensional, local-scale approach. *Geog Anal* 2016;**48**(2):132–56.
23. Prud'homme R, Lee G. Sprawl, speed and the efficiency of cities. *Urban Stud* 1999;**36**: 1849–58.
24. Richardson HW, Chang-Hee CB. *Urban sprawl in Western Europe and the United States.* London: Ashgate; 2004.
25. Scott AJ. *Global city-regions-trends, theory, policy.* New York: Oxford Unversity Press; 2001.
26. Scheyer JM, Hipple KW. *Urban soil primer.* Lincoln, Nebraska: United States Department of Agriculture, Natural Resources Conservation Service, National Soil Survey Center; 2005. http://soils.usda.gov/use.
27. Burghardt W, Morel JL, Zhang G-L. Development of the soil research about urban, industrial, traffic, mining and military areas (SUITMA). *J Soil Sci Plant Nutr* 2015; **61**(1):3–21.
28. Jiang R, Wang M, Chen W, Li X. Ecological risk evaluation of combined pollution of herbicide siduron and heavy metals in soils. *Sci Total Environ* 2018;**626**:1047–56.
29. Pouyat RV, Szlavecz K, Yesilonis ID, Groffman PM, Schwarz K. Chemical, physical, and biological characteristics of urban soils. *Am Soc Agron* 2010;**55**:119–56.
30. Levin JM, Kim K-HJ, Moel JL, Burghardt W, Charzynski P, Shaw RK. *Soils within Cities: global approaches to their sustainable management.* Stuttgart: Catena-Schweizerbart; 2017.
31. WRB: World Reference Base for Soil Resources. *A framework for international classification, correlation and Communication.* FAO; 2006. p. 145.
32. Craul PJ. A description of urban soils and their desired characteristics. *J Arboriculture* 1985;**11**(11):330–9.
33. Konstantinova E, Minkina T, Sushkova S, Konstantinov A, Rajput VD, Sherstnev A. Urban soil geochemistry of an intensively developing Siberian city: a case study of Tyumen, Russia. *J Environ Manage* 2019;**239**:366–75.
34. Pavao-Zuckerman MA. The nature of urban soils and their role in ecological restoration in cities. *Restor Ecol* 2008;**16**(4):642–9.
35. MEA: Millennium Ecosystem Assessment. *Ecosystems and human well-being: synthesis.* Washington (DC): Island Press; 2005.
36. Anne B, Geoffroy S, Cherel J, Warot G, Marie S, Noël CJ, Louis MJ, Christophe S. Towards an operational methodology to optimize ecosystem services provided by urban soils. *Landscape Urban Plann* 2018;**176**:1–9.
37. Tresch S, Frey D, Bayon R-CL, Zanetta A, Rasche F, Fliessbach A, Moretti M. Litter decomposition driven by soil fauna, plant diversity and soil management in urban gardens. *Sci Total Environ* 2019;**658**:1614–29.
38. Google Earth, 2016. [Accessed 15 May 2019].
39. Kibblewhite MG. Contamination of agricultural soil by urban and peri-urban highways: an overlooked priority? *Environ Pollut* 2018;**242**:1331–6.
40. Kumar K, Hundal LS. Soil in the city: Sustainably improving urban soils. *J Environ Qual* 2016;**45**:2–8.
41. Ferreira CSS, Walsh RPD, Steenhuis TS, Shakesby RA, Nunes JPN, Coelho COA, Ferreira AJD. Spatiotemporal variability of hydrologic soil properties and the implications for overland flow and land management in a peri-urban Mediterranean catchment. *J Hydrol* 2015;**525**:249–63.
42. Kalantari Z, Ferreira CSS, Walsh RPD, Ferreira AJD, Destouni G. Urbanization development under climate change: hydrological responses in a peri-urban Mediterranean catchment. *Land Degrad Dev* 2017;**28**(7):2207–21.

43. Palla A, Gnecco I. Hydrologic modeling of Low Impact Development systems at the urban catchment scale. *J Hydro* 2015;**528**:361–8.
44. Leitão IA, Ferreira CSS, Ferreira AJD. Assessing long-term changes in potential ecosystem services of a peri-urbanizing Mediterranean catchment. *Sci Total Environ* 2019;**660**:993–1003.
45. Stolte J, Tesfai M, Øygarden L, Kværnø S, Keizer J, Verheijen F, Panagos P, Ballabio C, Hessel R. *Soil threats in Europe: status, methods, drivers and effects on ecosystem services*. European Union: JRC Technical Reports; 2015.
46. Ferreira CSS, Walsh RPD, Ferreira AJD. Degradation in urban areas. *Curr Opin Environ Sci Health* 2018a;**5**:19–25.
47. Jato-Espino. Spatiotemporal statistical analysis of the urban heat island effect in a Mediterranean region. *Sustainable Cities Soc* 2019;**46**:101427.
48. Ferreira CSS, Walsh RPD, Steenhuis TS, Ferreira AJD. Effect of peri-urban development and Lithology on streamflow in a Mediterranean catchment. *Land Degrad Dev* 2018b;**29**:1141–53.
49. Kodikara J, Islam T, Sounthararajah. Review of soil compaction: history and recent developments. *Transp Geotech* 2018;**17**:24–34.
50. Craul PJ, Klein CF. Characterization of streetside soils in Syracuse, New York. *METRIA* 1980;**3**:88–101.
51. EPA: Environmental Protection Agency. Evaluation of urban soils: suitability for green infrastructure or urban agriculture. United States Environmental Protection Agency 2011; Publication No. 905R1103.
52. Miller JD, Brewer T. Refining flood estimation in urbanized catchments using landscape metrics. *Landscape Urban Plann* 2018;**175**:34–49.
53. Lu Q, Gao Z, Ning J, Bi X, Wang Q. Impact of progressive urbanization and changing cropping systems on soil erosion and net primary production. *Ecol Eng* 2015;**75**:187–94.
54. Mehri A, Salmanmahiny A, Tabrizi ARM, Mirkarimi SH, Sadoddin A. Investigation of likely effects of land use planning on reduction of soil erosion rate in river basins: case study of the Gharesoo River Basin. *Catena* 2018;**167**:116–29.
55. Mushi CA, Ndomba PM, Trigg MA, Tshimanga RM, Mtalo F. Assessment of basin-scale soil erosion within the Congo River Basin: a review. *Catena* 2019;**178**:64–76.
56. Ferreira CSS, Walsh RPD, Blake WH, Kikuchi R, Ferreira AJD. Temporal dynamics of sediment sources in an urbanizing Mediterranean catchment. *Land Degrad Dev* 2017; **28**(8):2354–69.
57. EPA: Environmental Protection Agency. *Guidance Specifying management measures for sources of Nonpoint pollution in coastal waters - Chapter 4. Management measures for urban areas*. United States Environmental Protection Agency; 1993. EPA-840-B-92-002.
58. Line DE, White NM. Effects of development on runoff and pollutant export. *Water Environ Res* 2007;**79**:185–90.
59. Line DE, White NM, Osmond DL, Jennings GD, Mojonnier CB. Pollutant export from various land uses in the upper Neuse River Basin. *Water Environ Res* 2002;**74**: 100–8.
60. Ferreira CSS, Ferreira AJD, Pato RL, Magalhães MC, Coelho CO, Santos C. Rainfall-runoff-erosion relationships study for different land uses, in a sub-urban area. *Z Geomorphol* 2012;**56**(X):005–20.
61. Russel KL, Vietz GJ, Fletcher TD. Global sediment yields from urban and urbanizing watersheds. *Earth Sci Rev* 2017;**168**:73–80.
62. Goonetilleke A, Thomas E, Ginn S, Gilbert D. Understanding the role of land use in urban stormwater quality management. *J Environ Manag* 2005;**74**:31–42.

63. Yu S, Wu Q, Li Q, Gao J, Lin Q, Xu Q, Wu S. Anthropogenic land uses elevate metal levels in stream water in an urbanizing watershed. *Sci Total Environ* 2014;**488–489**:61–9.
64. Cachada A, Coelho C, Gavina A, Dias AC, Patinha C, Reis AP, da Silva EF, Duarte AC, Pereira R. Availability of polycyclic aromatic hydrocarbons to earthworms in urban soils and its implications for risk assessment. *Chemosphere* 2018;**191**:196–203.
65. Wu S, Zhou S, Bao H, Chen D, Wang C, Li B, Tong G, Yuan Y, Xu B. Improving risk management by using the spatial interaction relationship of heavy metals and PAHs in urban soil. *J Hazard Mater* 2019;**364**:108–16.
66. Keshavarzi B, Najmeddin A, moor F, Moghaddam PA. Risk-based assessment of soil pollution by potentially toxic elements in the industrialized urban and peri-urban areas of Ahvaz metropolis, southwest of Iran. *Ecotoxiol Environ Saf* 2019;**167**:365–75.
67. Barsova N, Yakimenko O, Tolpeshta I, Motuzova G. Current state and dynamics of heavy metal soil pollution in Russian Federation: a review. *Environ Pollut* 2019;**249**: 200–7.
68. Li C, Sun G, Wu Z, Zhong H, Wang R, Liu X, Guo Z, Cheng J. Soil physiochemical properties and landscape patterns control trace metal contamination at the urban-rural interface in southern China. *Environ Pollut* 2019;**250**:537–45.
69. Kuusisto P, Hjort J. Land use impacts on trace metal concentrations of suburban stream sediments in the Helsinki region, Finland. *Sci Total Environ* 2013;**456–457**:222–30.
70. Pitt RE, Maestre A. *Stormwater quality as described in the National stormwater quality Database (NSQD).* 10th International Conference on Urban Drainage, Copenhagen/ Denmark. 2005.
71. Walsh D, Glass K, Morris S, Zhang H, RcRae I, Anderson N, Alfieri A, Egendorf SP, Holberton S, Owrang S, Cheng Z. Sediment exchange to mitigate pollutant exposure in urban soil. *J Environ Manage* 2018;**214**:354–61.
72. Brtnický M, Pecina V, Hladký J, Radziemska M, Koudelková Z, Klimánek M, Richtera L, Adamcová D, Elbl J, Galiová MV, Baláková L, Kynický J, Smolíková V, Houska J, Vaverková MD. Assessment of phytotoxicity, environmental and health risks of historical urban park soils. *Chemosphere* 2019;**220**:678–86.
73. Kikuchi R, Ferreira CSS, Kanoun-Boulé M. The issues of pet waste (canine feces) – environmental problems, human behavior and knowledge of public health. *Int J Interdiscip Res Innovations* 2017;**5**(1):34–42.
74. Kikuchi R, Ferreira CS, Ferreira AD. Environmental survey of heavy metals in the Yiluo Basin of East-central China - water, soil and plants. *World J Res Rev* 2016; **3**(1):19–23.
75. EP: European Parliament. *Directive 2010/75/EU on industrial emissions (integrated pollution prevention and control).* 2010. L 334/17.
76. Kulabako NR, Nalubega M, Thunvik R. Study of the impact of land use and hydrogeological settings on the shallow groundwater quality in a peri-urban area of Kampala, Uganda. *Sci Total Environ* 2007;**381**:180–99.
77. Azzellino A, Colombo L, Lombi S, Marchesi V, Piana A, Andrea M, Alberti L. Groundwater diffuse pollution in functional urban areas: the need to define anthropogenic diffuse pollution background levels. *Sci Total Environ* 2019;**656**:1207–22.
78. Steffy LY, Kilham SS. Elevated d15N in stream biota in areas with septic tank systems in an urban watershed. *Ecol Appl* 2004;**14**:637–41.
79. Gold AJ, DeRagoon WR, Sullivan WM, Lemunyon JL. Nitrate-nitrogen losses to groundwater from rural and suburban land uses. *J Soil Water Conserv* 1990;**45**:305–10.
80. Kosheleva NE, Vlasov DV, Korlyakov ID, Kasimov NS. Contamination of urban soils with heavy metals in Moscow as affected by building development. *Sci Total Environ* 2018;**636**:854–63.

81. Emmenegger L, Mohn J, Sigrist M, Marinov D, Steinemann U, Zumsteg F, Meier M. Measurement of ammonia emissions using various techniques in a comparative tunnel study. *Int J Environ Pollut* 2004;**22**:326–40.
82. Soonthornnonda P, Christensen ER, Liu Y, Li J. The sources, impact and management of car park runoff pollution: a review. *Sci Total Environ* 2008;**402**:248–56.
83. Almakki A, Jumas-Bilak E, Marchandin H, Licznar-Fajardo P. Antibiotic resistance in urban runoff. *Sci Total Environ* 2019;**667**:64–76.
84. Gromaire MC, Garnaud S, Saad M, Chebbo G. Contribution of different sources to the pollution of wet weather flows in combined sewers. *Water Res* 2001;**35**(2):521–33.
85. Good JC. Roof runoff as a diffuse source of metals and aquatic toxicity in stormwater. *Water Sci Technol* 1993;**28**(3–5):317–21.
86. Simmons G, Hope V, Lewis G, Whitmore J, Gao W. Contamination of potable roof-collected rainwater in Auckland, New Zealand. *Water Res* 2001;**35**(6):1518–24.
87. Steuer J, Selbig W, Hornewer N, Prey J. *Sources of contamination in an urban basin in Marquette, Michigan and an analysis of concentrations, loads, and data quality*. Water Resources Investigations Report 97-4242. Middleton, Wisconsin: U.S. Geological Survey and U.S. Environmental Protection Agency; 1997.
88. Liu Y, Gao P, Su J, da Silva EB, de Oliveira LM, Townsend T, Xiang P, Ma LQ. PAHs in urban soils of two Florida cities: background concentrations, distribution, and sources. *Chemosphere* 2019;**214**:220–7.
89. Korashy HM, El-Kadi AOS. The role of Aryl hydrocarbon Receptor in the Pathogenesis of cardiovascular diseases. *Drug Metab Rev* 2006;**38**(3):411–50.
90. ATSDR – Agency for Toxic Substances and Disease Registry. The ATSDR 2017 Substance Priority List. http://www.atsdr.cdc.gov/spl/.
91. Dias-Ferreira C, Pato RL, Varejão JB, Tavares AO, Ferreira AJD. Heavy metal and PCB spatial distribution pattern in sediments within an urban catchment—contribution of historical pollution sources. *J Soils Sediments* 2016;**16**(11):2594–605.
92. NTP: National Toxicology Program. *Report on carcinogens*. 14th ed. New York: U.S. Department of Health and Human Services, Public Health Service; 2016.
93. Ettema CH, Wardle DA. Spatial soil ecology. *Tends Ecol Evol* 2002;**17**:117–83.
94. Parkin TB. Spatial variability of microbial processes in soil – a review. *J Enviorn Qual* 1993;**22**:409–17.
95. Bloem J, Benedetti A, Hopkins D. *Microbial methods assessing soil quality*. CABI Publishing; 2006.
96. Kennedy AC, Smith KL. Soil microbial diversity and the sustainability of agricultural soils. *Plant Soil* 1995;**170**:75–86.
97. Agarwal L, Qureshi A, Kalia VC, Kapley A, Purohit HJ, Singh RN. Arid ecosystem: future option for carbon sinks using microbial community intelligence. *Curr Sci* 2014; **106**:10.
98. Mahnert A, Moissl-Eichinger C, Zojer M, Bogumil D, Mizrahi I, Rattei T, Martinez JL, Berg G. Man-made microbial resistances in built environments. *Nat Commun* 2019;**10**:968.
99. Reese AT, Savage A, Youngsteadt E, McGuire KL, Koling A, Watkins O, Frank SD, Dunn RR. Urban stress is associated with variation in microbial species composition—but not richness—in Manhattan. *ISME J* 2015;**10**:751–60.
100. Hanski I, von Hertzen L, Fyhrquist N, Koskinen K, Torppa K, Laatikainen T, Karisola P, Auvinen P, Paulin L, Makela MJ, Vartianinen E, Kosunen TU, Alenius H, Haahtela T. Environmental biodiversity, human microbiota, and allergy are interrelated. *Proc Natl Acad Sci* 2012;**109**:8334–9.
101. Kowalchuk GA, Bruinsma M, Van Veen JA. Assessing responses of soil microorganisms to GM plants. *Trends Ecol Evol* 2003;**18**:403–10.

102. Ramirez KS, Leff JW, Barberán A, Bates ST, Betley J, Crowther TW, Kelly EF, Oldfield EE, Shaw EA, Steenbock C, Bradford MA, Wall DH, Fierer N. Biogeographic patterns in below-ground diversity in New York City's Central Park are similar to those observed globally. *Proc R Soc B* 2014;**281**:20141988.
103. Xu HJ, Li S, Su JQ, Nie S, Gibson V, Li H, Zhu YG. Does urbanization shape bacterial community composition in urban park soils? A case study in 16 representative Chinese cities based on the pyrosequencing method. *FEMS Microbiol Ecol* 2014;**87**:182–92.
104. Yan B, Li J, Xiao N, Qi Y, Fu G, Liu G, Qiao M. Urban-development-induced changes in the diversity and composition of the soil bacterial community in Beijing. *Sci Rep* 2016;**6**:38811.
105. Lundy L, Wade R. Integrating sciences to sustain urban ecosystem services. *Prog Phys Geogr* 2011;**35**(5):653–69.
106. Sarwar N, Imran M, Shaheen MR, Ishaq W, Kamran A, Matloob A, Rehim A, Hussain S. Phytoremediation strategies for soils contaminated with heavy metals: Modifications and future perspectives. *Chemosphere* 2017;**171**:710–21.
107. Goldenberg R, Kalantari Z, Destouni G. Increased access to nearby green–blue areas associated with greater metropolitan population well-being. *Land Degrad Dev* 2018;**29**:3607–16.
108. Kabisch N, Frantzeskaki N, Pauleit S, Naumann S, Davis M, Artmann M, Haase D, Knapp S, Korn H, Stadler J, Zaunberger K, Bonn A. Nature-based solutions to climate change mitigation and adaptation in urban areas: perspectives on indicators, knowledge gaps, barriers, and opportunities for action. *Ecol Soc* 2016;**21**(2):39.
109. Maes J, Jacobs S. Nature-based solutions for Europe's sustainable development. *Conserv Lett* 2017;**10**(1):121–4.
110. Kabisch N, van den Bosch MA. Urban green spaces and the potential for health Improvement and environmental Justice in a changing climate. In: Kabisch N, Korn H, Stadler J, Bonn A, editors. *Nature-based solutions to climate change adaptation in urban areas: Linkages between Science, policy and practice*. Cham: Springer International Publishing; 2017. p. 207–20.
111. Elmqvist T, Fragkias M, Goodness J, Güneralp B, Marcotullio PJ, McDonald RI, Parnell S, Schewenius M, Seto MS, Wilkinson KC. *Urbanization, biodiversity, and ecosystem services: challenges and opportunities*. New York: Springer; 2013.
112. Pan H, Deal B, Destouni G, Zhang Y, Kalantari Z. Sociohydrology modeling for complex urban environments in support of integrated land and water resource management practices. *Land Degrad Dev* 2018;**29**:3639–52.
113. EC. *Best practices on flood prevention, protection and mitigation*. 2003. Water Directors meeting in Athens in June 2003.
114. Cohen-Shacham E, Walters G, Janzen C, Maginnis S. *Nature-based solutions to address societal challenges*. Gland: IUCN; 2016.
115. Gupta AK, Nair SS. *Ecosystem approach to disaster risk reduction*. New Delhi, India: National Institute of Disaster Management; 2012. p. 202.
116. UNEP 2019, UN Decade on Ecosystem Restoration: https://www.unenvironment.org/news-and-stories/press-release/new-un-decade-ecosystem-restoration-offers-unparalleled-opportunity.
117. SER International Primer on Ecological Restoration. Society for Ecological Restoration: 2004. Available from: URL: http://www.ser.org/resources/resourcesdetail-view/ser-international-primer-onecological-restoration.
118. EC. *Communication from the Commission to the European Parliament, the Council, the European economic and social Committee and the Committee of the regions: green infrastructure (GI) – enhancing Europe's natural capital*. COM 249 final, Brussels. 2013.
119. Potschin M, Kretsch C, Haines-young R, Furman E, Berry P, Baró F. *Nature-based solutions*. OpenNESS 1604 Ecosystem Service Reference Book; 2016.

120. Keesstra SD, Bouma J, Wallinga J, Tittonell P, Smith P, Cerdà A, Montanarella L, Quinton JN, Pachepsky Y, van der Putten WH, Bardgett RD, Moolenaar S, Mol G, Jansen B, Fresco LO. The significance of soils and soil science towards realization of the United Nations Sustainable Development Goals. *Soil* 2016;**2**:111–28.
121. Davies J. Enabling governance for sustainable land management. In: Chabay I, Frick M, Helgeson J, editors. *Land restoration: reclaiming landscapes for a sustainable future*. Massachusetts: Academic Press, Elsevier, Cambridge; 2015. p. 598.
122. Fernandes JP, Guiomar N. Nature-based solutions: the need to increase the knowledge on their potentialities and limits. *Land Degrad Dev* 2018;**29**:1925–39.
123. Thorslund J, Cohen MJ, Jawitz JW, Destouni G, Creed IF, Rains MC, Badiou P, Jarsjö J. Solute evidence for hydrological connectivity of geographically isolated wetlands. *Land Degrad Dev* 2018;**29**:3954–62.
124. Rodrigo-Comino J, Giménez-Morera A, Panagos P, Pourghasemi HR, Pulido M, Cerdà A. The potential of straw mulch as a nature-based solution for soil erosion in olive plantation treated with glyphosate: a biophysical and socioeconomic assessment. *Land Degrad Dev* 2019;**1**– **13**.
125. Pan D, Gao X, Huang J, Yang M, Wu P, Wang J, Dyck M, Zhao X. Vegetative filter strips - effect of vegetation type and shape of strip on runoff and sediment trapping. *Land Degrad Dev* 2018:1–11.
126. Hellies M, Deidda R, Viola F. Retention performances of green roofs worldwide at different time scales. *Land Degrad Dev* 2018;**29**:1940–52.
127. Peh KSH, Balmford A, Field RH, Lamb A, Birch JC, Bradbury RB, et al. Benefits and costs of ecological restoration: rapid assessment of changing ecosystem service values at a U.K. wetland. *Ecol Evol* 2014;**4**:3875–86.
128. EEA: European Environment Agency. *Flood risk and environmental vulnerability: Exploring the synergies between floodplain and restoration, water policies and thematic policies*. EEA report No.1/2016. Copenhagen. Denmark. 2016.
129. Short C, Clarke L, Carnelli F, Uttley C, Smith B. Capturing the multiple benefits associated with nature-based solutions: lessons from a natural flood management project in the Cotswolds, UK. *Land Degrad Dev* 2019;**30**:241–52.
130. ISDR. *Hyogo framework for action 2005-2015*. Building the Resilience of Nations and Communities to Disasters Extract from the final report of the World Conference on Disaster Reduction (A/CONF.206/6). 2005.

9780128164150